CW00954454

STAR THEATRE

STAR THEATRE

The Story of the Planetarium

William Firebrace

REAKTION BOOKS

Published by Reaktion Books Ltd
Unit 32, Waterside
44–48 Wharf Road
London N1 7UX, UK
www.reaktionbooks.co.uk

First published 2017

Printed and bound in China by 1010 Printing International Ltd

A catalogue record for this book is available from the British Library

ISBN 978 1 78023 835 7

CONTENTS

MISSING PLANET

A diminutive planet, mounted on a slender pole, hovers above the London Planetarium. This planet is whitish in colour, about 1 metre in diameter, and encircled by a flat, white disc, representing a ring of cosmic dust. The planetarium closed some years ago, under never quite explained circumstances, and now functions merely as a 3D cinema attached to the neighbouring Madame Tussauds waxworks museum. At night, however, up in the sky, the little planet still glows softly with a yellowish light, quietly ignoring the hubbub of traffic on the road below.

I used to work in an office with a window facing the London Planetarium, and found this planet rather reassuring, a sign that all was well in the solar system. So when the planet suddenly vanished, sometime during the summer of 2012, I was gripped by feelings of worry and anxiety. Planets are meant to be reasonably constant in their elliptical paths around the Sun, a reassuring celestial register of the stability of the solar system as a whole, including our own circulating planet.

The case of the missing planet. Sherlock Holmes, consulting detective resident in nearby 221B Baker Street, would perhaps have been intrigued. His nemesis, Professor James Moriarty, the 'Napoleon of crime', was well known for his volume *The Dynamics of an Asteroid*, which coincidentally explored a missing planet

London Planetarium with its planet-topped dome.

located somewhere between Mars and Jupiter. A case in a detective story often starts with a simple clue – a disappearance or unexpected event – but leads to a mass of individual, entangled threads, sometimes without any obvious means of resolution. Usually the original clue becomes irrelevant as more complex matters evolve.

The disappearance of the small planet was a sign of some greater absence, suggesting certain questions, about not just the London Planetarium but planetariums in general. Planetariums are part of most people's childhood experience, usually dimly remembered from a school visit and recalled by adults paying a return visit with their own children. These questions begin as simple wondering, and rapidly become more complex. How and where did the planetarium originate? What kind of simulation of the solar system and the universe does the planetarium produce? How does the planetarium mix theatre with science? How has the planetarium changed with developments in astronomy? What is the relationship between the exterior and interior of the building?

In this book, these and other themes will be investigated over five approximately chronological sections: the precedents for the planetarium; its invention in Germany in the 1920s; developments in the USSR and the U.S. in the 1930s; the expansion across the globe in the later twentieth century, at the time of the space race; and finally, the evolution of the contemporary planetarium in our own period of startling astronomical and cosmological discovery. But the planetarium does not simply follow a line in time: it slips back and forth, approaching and receding, like some errant celestial body.

The London Planetarium planet reappeared on the roof of the dome in December 2012, returning as unexpectedly as it vanished. A group of building workers crawled up the ladder on the side of the dome, passing between them with considerable effort the globe, which they reattached to the pole on the dome. This pole has an ingenious mechanism allowing it to be lowered to a horizontal position, making the task easier. Some days later the planet was once again glowing, restoring a sense of reliability and comfort. According to the owners of Madame Tussauds, the planet had been 'undergoing standard maintenance'. Cosmic harmony has been restored – at least for now. The links between astronomy and architecture continue to evolve in unexpected ways within the star theatre.

Zeiss planetary projection,
Jena, 1920s.

ONE

HOLY, ROUGH, IMMEDIATE

'The sun shone,' begins Samuel Beckett's novel *Murphy* (1938), 'having no alternative, on the nothing new.' The planetarium produces an illusion of the stars and planets at night rather than the Sun during the day, but what it offers is a similarly repetitive view of the solar system: the Sun sets in the west to reveal the same stars in the night sky, the same planets follow their designated paths, and the Sun rises again in the east as the stars fade away. The spectacle returns for each performance; the traditional planetarium is the reliable but always slightly surprising theatre of the night.

Planetarium shows occur not in some arid classroom or lecture hall, but in a place of entertainment, producing a mechanical performance in which lights shine and move across the dome, where popcorn and ice cream are available, and where emotion counts as much as intellect. The display is intended for all, but is primarily attended by children, who are given a mix of science and thrills. Most people can recall a childhood visit to a planetarium: the unusual environment, the sense of anticipation, the room darkening, the stars coming up, the movement of the lights, the voice of the lecturer. Many children visit a planetarium only once, but this can become a defining moment, as the night sky is revealed for the first time – a moment that remains in the memory, and which can surface unexpectedly.

Peter Harrison Planetarium, Greenwich.

The familiar, usually domed building within which the solar system is projected in shows put on for scientific education and entertainment is to be found today in most cities. The planetarium projector and dome were invented in the early 1920s in the German city of Jena, by the engineer Walther Bauersfeld, working for the firm Carl Zeiss, an important manufacturer of optical instruments. In essence, in a planetarium a complex set of projectors beams moving lights, representing the planets and stars, onto a large, hemispherical screen. The audience, who as inhabitants of the modern city often have little idea of the appearance of the night sky, are seated in a circle around the projector, gazing upwards at the movements of the artificial sky. Today, almost a century since its invention, the planetarium still functions in a very similar fashion to the original design, but has been considerably adapted according to both the technological improvements to the projection system and advances in astronomy, which offer a completely different version of the universe to that which was understood a century ago.

THE PLANETARIUM WAS ORIGINALLY KNOWN, in the 1920s, as the star theatre. It produces a live show, directed and commented on by the lecturer – certainly a very individual kind of theatre. Theatre comes in any number of forms: traditional, realist, absurd, surreal, expressionist, musical, kitchen sink, heroic, tragic and so on. The English avant-garde theatre director Peter Brook defined four kinds of theatre in his book *The Empty Space* (1968): deadly (traditional and for the most part boring and thus to be avoided), holy (or the 'invisible-made-visible'), rough (unsophisticated, popular) and immediate (direct, with minimum gap between audience and performer). These types of theatre are not always completely separated, sometimes 'mixing together within one performance', which can have aspects of holy, rough and immediate. For Brook these categories form a basis for considering the different ways plays are put on and performed. So, what kind of theatre is the star theatre? It seems to be realist, or naturalistic, since it shows a visual simulation of the night sky. On occasions it can also be deadly and predictable – one can only be surprised to hear that the universe is very large so many times. But, as will be seen in this book, its antecedents lie also in the holy, the rough and the immediate – and other types of performance.

Early visitors to the planetarium were entranced, sometimes even imagining that somehow the dome had opened and that they were looking at the real sky. Keeping some element of this naivety is still part of the charm of the show. The star theatre of Walther Bauersfeld does two different things, ingeniously entwined so that the spectator does not distinguish between the two: first, it produces a simulation of the view one might see in the sky on a clear night – effectively, an artificial imitation of nature. Second, it illustrates a theory of how the planets move around the Sun, and how the solar system relates to the stars and the universe beyond. A view – how we see the night sky – is shown together with a model – how the solar system, the stars and the universe are thought to function.

The two do not necessarily coincide. There is a difference between what we see and what is actually there. It is quite possible to look at the actual night sky and to have any number of ideas about what is going on in it, as shown by the numerous different opinions on the structure of the solar system throughout history. It would be very natural to consider, as one stands on the surface of the Earth, that one is at the centre of movement, rather than on just another planet rotating around one of many stars. Furthermore, there is not only one 'up'; one's view of the stars varies according to the location on the surface of the Earth where one is standing, so that a viewer in the southern hemisphere will see completely different stars to a viewer in the north. The move away from considering the Earth as the centre of the solar system and of the universe is part of a much larger development – a gradual awareness that we are just one of many species living on a planet revolving around one of countless stars.

Today our understanding of the size of the universe has grown since the more limited astronomy of the 1920s – the Sun as a star in the Milky Way galaxy, which in turn is one of an uncountable number of galaxies expanding rapidly away from one another into a universe of undefined, perhaps indefinable, size. This expansion into the outer reaches of space naturally reaches beyond the planets and stars we are able to see in the sky at night – far beyond what anyone can see from the surface of the Earth without the use of a very high-resolution telescope. The traditional view of the night sky, as projected by the early planetariums, becomes a minute part of a much greater universe. As the fox in Antoine de Saint-Exupéry's tale *The Little Prince* puts it, 'One sees clearly only with the heart. Anything essential is invisible to the eyes.'

Beckett's 'nothing new' is a world-weary novelistic assumption, suitable to his own melancholic view of the world but not to a description derived from astronomy. The Sun always shines on something new, and there are always new theories of the circumstances under which it shines. The planetarium offers only the most convincing of a sequence of notions of the relationship of the Earth to the solar system and to the surrounding universe. There have been in the past any number of theories of just what the view of the sky implies, each linked to religion notions, observational abilities, mathematical knowledge and cultural inclinations. Many of these propositions involved a greater use of imagination than observation. The Earth has, in an absurd and possibly mythical example, been proposed as a flat plate sitting on a stack of turtles – the turtles, according to the woman who supposedly offered this ancient theory to a baffled Bertrand Russell at one of his lectures, going 'all the way down'. The question of where the turtles go down to would always be an awkward one, for what lies beyond the last visible turtle? Somewhere, there may well be a planetarium illustrating this intriguing theory.

The universe, according to Douglas Adams's skit on contemporary cosmology, *The Hitchhiker's Guide to the Galaxy* (1979), is an amateur construction put together by an old man and a cat, the old man creating and selling luxury planets. The Earth is one of his creations – the cosmos as a commercial product. Unfortunately, the Earth is demolished to construct an intergalactic highway – an appropriate enough theory for a period where anything can be made and sold.

Every theory requires some kind of model to help explain it to others and convince them, in the absence of any obvious clues in the visible night sky, of its truth. This model might take a variety of forms – a wall painting, a machine, a painted dome, a projection system, a computer animation – according to the requirements of the theory and the custom of the time. The evolution of such models does not follow a straight line, but leads gradually and with many enjoyable detours, dead ends, short-cuts and still unexplored paths to the invention of the planetarium and its demonstration of the solar system. The detours and dead ends are important; sometimes they influence at a later date the models' main path of evolution, and besides, they are often more fun than the main path. What seems at first a dead end can unexpectedly reveal a way forward. Contemporary models will no doubt at some time in the future be equally outmoded.

What might be specific examples of the early theatres of the stars, which lead up to the invention of the true planetarium? There is only space here to examine a few of the many which existed; we will concentrate mostly on those which relate to the scale of the planetarium as a theatre offering a show, rather than merely astronomical machines and devices. One can start with Peter Brook's categories of the holy, the rough and the intermediate, and find also appearing alongside these the royal, the mechanical and the spherical.

The Holy Theatre I

Older theories of the solar system may now appear naive, but they were always appropriate to the time during which they were formed. Traces of these theories survive unexpectedly in how we think today. The ancient Egyptians considered the sky to be a flat ceiling positioned over a flat Earth. The goddess Nut, her body studied with stars, was believed to swallow the Sun each evening and give birth to it again in the morning, while the Sun god Ra travelled each day across the sky, descending in the west

The Egyptian goddess Nut, her body a starry sky enclosing the Earth.

into the underworld, where he encountered its ruler Osiris, fought with the serpent Apep, to return on time in the morning. The reappearance of the Sun in the east was linked to the expected rebirth of the human soul, astronomy emerging from spiritual belief. The Egyptians never conceived of the skies as a dome; since their world was based on the long valley of the Nile, with deserts either side, their image of the stars was linked to ideas of reincarnation and had little relationship to a three-dimensional night sky. They produced wonderful two-dimensional wall and sarcophagus paintings showing the body of Nut, covered in stars, bent over and covering the Earth, with various other smaller figures gazing upwards: a goddess as the earliest form of planetarium. The same repeated divine routine – an early version of Beckett's solar routine – was required to keep the stars appearing at night and the Sun returning each morning. There is lurking here somewhere the idea that the rite needed to be performed to ensure the return of the Sun. If the god failed to arrive on time, the daily life of the known world would come to a halt.

Royal Theatre

But what if, rather than looking at a two-dimensional image of the skies, the spectator is placed within a three-dimensional space, providing an illusion of a three-dimensional version of the heavens – as occurs within the planetarium? This idea was developed surprisingly early. The Sasanian king Khosrow II, who reigned in the seventh century AD in what is now southern Iran, just before the arrival of Islam, desired to demonstrate that he ruled not only the land but the heavens. He had constructed a palace with a domed throne room, and the interior of the dome was decorated with the Moon, the stars, the planets and the figures of the zodiac, the astrological signs that also form the astronomical divisions of celestial longitude. The historian Matthew Canepa writes in *The Two Eyes of the Earth* (2009),

> The domed interior of the throne's baldachin, which covered the king of kings' place, displayed a lapis lazuli sky with the stars of the zodiac, the planets, seven *kišwarān* [divisions of the heavens], and the Sun and the Moon in precious metals and jewels. This domed canopy rotated in concert with the movement of the heavens such that one could tell time by its movement, and creating the illusion that heaven and Earth rotated around the king . . . In addition to mimicking the movement of heavens, the throne's attributes and the king's place on it changed with the passage of the seasons . . . four jewel-encrusted carpets were alternately laid on the throne's lower surfaces, one for each season.

The Khosrow dome was an early star theatre, using elaborate statecraft to show that Khosrow was a greater king than any of his rivals. The theatre was effective so long as the machinery continued to perform; if it malfunctioned the movements of the artificial sky came to a halt. The dome of King Khosrow, although not scientific, already sets out the basic principles of both the contemporary planetarium, in which the movements of the solar system rotate around the spectator, though today the projector has replaced the king at the centre.

The Khosrow dome belongs to a now distant time, but it has modern-day equivalents in its desire to create a dome of the sky in order to exert control. For instance, the artificial illusion of the sky which needs to be continually maintained in order that daily life functions finds its contemporary form in the film *The Truman Show* (1998). An insurance agent, played by Jim Carrey, has lived his entire life in an elaborately constructed simulated world created for a television series. An artificial sky with its own weather system, set within an enormous dome and run by the TV company, covers his town. 'Cue the Sun,' commands the show's director, hidden in the control room beyond the dome, and the Sun comes out. Eventually the artificial sky malfunctions, and the Sun emerges at the same time as the Moon; the illusion collapses and the leading character is freed. In the Khosrow example, the king controls the heavens in order to display his authority. In the film, it is the media empires that dictate the movements of the skies, for the purpose of putting on a daily TV show. The stagecraft has become more sophisticated, but the aim of both king and television studio is similar: control the skies, and one controls also the lives of one's subjects.

The Spherical Theatre I

The heavens were imagined in medieval times in Europe as a finite series of spheres, like the layers of an onion, with the Earth at the centre, surrounded by the sequence of visible planets reaching out to the final spheres of the stars and of the prime mover, the being who makes all the others revolve. The sphere was considered a perfect form, and the universe, made by God, was necessarily perfect. Traces of this idea survive today in the hemispherical screen of the planetarium, a relic of a vanished ideal. Such theories of the solar system were illustrated both with complex drawings, showing the layers of circular movement, and later also with physical models called armillary spheres.

Spheres demonstrating the movement of the planets were constructed at a large, almost habitable scale, such as the great armillary sphere produced in 1588–93 for Ferdinando I de' Medici by the mathematician, cartographer and instrument maker Antonio Santucci delle Pomarance. Its wonderfully intricate construction can now be seen in the Museo Galileo in Florence. The Santucci sphere stands over 3 metres tall, an extraordinary construction of interlocking metal rings, all coated in gold leaf, dwarfing both the viewer and all the other spheres in the room – a true piece of cosmic machinery. It illustrates the Ptolemaic system, with the Earth at the centre, seven spheres for the Sun, Moon and wandering planets (Mercury, Venus, Mars, Jupiter, Saturn), a ring for fixed stars, and a ninth for the zodiac and finally the tenth sphere for the Primum Mobile, which puts all this complex celestial mechanics in motion. This sphere is based on the false premise that the Sun revolves around the Earth, but the skill of its construction is so astonishing, and the details so immaculate, that the viewer could be easily convinced that it is a reasonable explanation of how the solar system operates. The best model does not necessarily have to be based on the best theory. The viewer is mesmerized by the three-dimensional pattern of the rings, and drawn to the small wooden Earth at the centre, which represents his actual position. Such astronomical machines begin to cross over into habitable spaces: their size suggests that they could be lived in, and their relationship to the human body changes.

Santucci delle Pomarance's armillary sphere with the Sun, Moon, five planets, the fixed stars, the zodiac and the prime mover, 1593.

One might wonder again just what lies beyond the prime mover – who or what moves the mover – a matter to which numerous theologians have devoted considerable time, for the most part without reaching any useful conclusion. The shadowy figure of the prime mover who – or which – sets everything in motion, will reappear repeatedly, in divine and less divine forms, throughout the story of the planetarium. In another context, William Butler Yeats in his poem 'Meditations in Time of Civil War' (1922) would credit the prime mover with remarkable circular powers, linking the perfect movements of the planets to those of night creatures: "The Primum Mobile that fashioned us/ Has made the very owls in circles move.'

In medieval times, only rare philosophers, such as Nicholas of Cusa in his subversively named book *On Learned Ignorance* (1440), had considered that the universe might not be spherical and have no defined limits. It would be difficult to make a physical model of a universe which had no limits, but the matter has returned to the planetarium today as we again consider how to explain a universe with neither a clear centre nor limits, and which contains phenomena such as dark holes that defy physical representation.

When the mathematical, observation-based systems of Johannes Kepler and Isaac Newton replaced these comfortable nesting spheres of medieval astronomy, the planets acquired physical weight and followed their paths according to the laws of gravity. Astronomical models, however, remained delightfully imaginative, refusing to merely accept the consequences of scientific investigation. Kepler produced an intricate model of the solar system in the form of a drinking cup with the orbits of the visible planets – Mercury, Venus, Earth, Mars, Jupiter, Saturn – alternating with the five regular polyhedra – pyramid, cube, octahedron, dodecahedron and icosahedron – the one series set within the other. In the drinking cup, each planet poured out a different kind of wine. In fact, the wonderful system did not work, for Kepler later calculated that the planetary orbits do not follow a perfect circle but an ellipse. His interlocking planets and solids were a cul-de-sac in

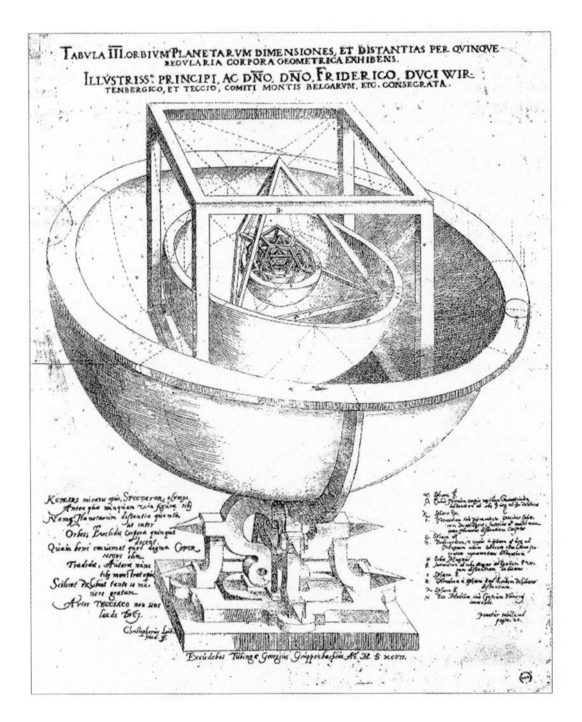

The cosmic cup of Johannes Kepler with the planetary spheres set within the Platonic solids.

terms of scientific progress, but the sphere within a geometrical solid reappears later – without, however, Kepler's mystic tendencies – in the construction of the first planetariums, which were domes within lattice structures. Kepler remained attached to a cosmos based on the projection of light. 'The perfection of the world', he wrote in *Epitome of Copernican Astronomy (1617–21)*,

> consists in light, heat, movement, and in harmony of movements . . . the sun is very beautiful with light and is as if the eye of the world, like a source of light or very brilliant torch, the sun illuminates, paints, and adorns the bodies of the rest of the world . . . the sphere of the fixed stars plays the role of the riverbed in which this river of light runs, and is as it were an opaque and illuminated wall, reflecting and doubling the light of the sun.

The Sun becomes a curious eye emitting light, a luminous prime mover. That the solar system is based on a system of light is the essential quality of the planetarium of the 1920s, with the projector replacing the Sun as the source of light.

The Mechanical Theatre I

The observations and calculations of just how the planets move required complicated miniature machines, which became known as orreries, showing the movement of the planets. The devising of these machines often involved intricate clockwork mechanisms, paralleling the development of astronomical ideas of the cosmos as an enormous clockwork machine, wound up at a time far back in the past and continuing to perform its mechanical ballet until the end of time.

But stars and planets moving around on rods propelled by clockwork are really not enough for an ambitious model of the sky.

Overleaf:
The astronomical clock and celestial globe by Schwilgué at Strasbourg Cathedral, 1838–43.

ASSYRIA
ASSOMPTION
Quatre Temps
LEVER DU SOLEIL
GRÆCIA

APPARENT
COUCHER DU SOLEIL
OCÉAN ATLANTIQUE
AMÉRIQUE
PAQUES
MARS
FÉVRIER
JANVIER
ROMA

Such a model needs a larger cast: it needs gods, animals, macabre scenes, sounds, competing theories of how time operates. We need a better clockmaker. We need Jean-Baptiste Schwilgué.

Schwilgué was born in Strasbourg in eastern France, but his family was evicted from the city for political reasons. In exile, he recalled the astronomical clock of Strasbourg Cathedral, whose machinery had long ago broken down. 'Inspired by the idea to bring back to life the masterpiece of his native town,' wrote his apprentice Alfred Ungerer in his pamphlet on the clockmaker,

> Schwilgué became obsessed with minute mechanisms. Alone, without a master, he perfected the art of clockmaking, with its complex mechanisms, and could create with extraordinary precision astronomical pendulums and counters for seconds . . . A true autodidact, he extended his knowledge of astronomy, mathematics, physics and mechanics through the books he acquired.

Schwilgué is of a type who returns repeatedly in the story of the planetarium: the self-taught obsessive who through his ingenuity creates a device that shows the movements of the solar system in a new way – in his case in combination with myths from various religions. Schwilgué returned to Strasbourg and rebuilt the city's ancient astronomical clock, but with extra dials and figures, creating a theatrical astronomical machine capable of putting on a performance on the nature of time. The massive clock, which stands 7 metres tall, combines dials and models showing the positions of the planets and moons with automata of the planetary gods on chariots, processions of religious figures, Jesus and the Apostles, and other figures showing human time – a child, an adult, an old person and a life-size crowing cockerel. There is a touch of Christian ritual here; the old and new gods need to appear, the dials need to keep time, not just to show what is happening in the solar system but to keep it turning.

The clock's individual elements are contained in a large case, almost a piece of architecture in itself. Time, the Schwilgué clock suggests, can move at many speeds, and also derive from different aspects of culture – scientific, everyday, spiritual – suggesting the existence of different kinds of time, a remarkably prescient idea considering the diverse theories of time proposed in our age, such as in Stephen Hawking's *A Brief History of Time* (1988). Today, even if we have discarded the figures of the old gods, we still struggle to produce a coherent view of the universe. And as with the Santucci sphere, the scale of the operation changes the relationship of the display to the viewer, who is among others, part of an audience captivated by the mechanical show producing for a few minutes every day a religious-astronomical performance. It is left uncertain just which are the most important elements – the scientific dials, or the mythical figures on their chariots.

The Mechanical Theatre II

The Strasbourg clock is a display for the public. But why not have one on the ceiling of one's living room? An extraordinary and rather eccentric model of the solar system was built in the years 1774–81 by the Dutchman Eise Eisinga, by profession a wool carder, another mechanical obsessive and an amateur astronomer, in the living room of his house in Franeker in the flat landscape of the northern coast of the Netherlands. Eisinga constructed his planetarium at a time when an unusual conjunction of the Moon with the planets Mercury, Venus and Mars was predicted for 8 May 1774. According to a local pastor, Eelco Alta, this conjunction had last taken place at the time of Creation, implying the impending end of the world, as the planets would collide and Earth be knocked off its orbit into

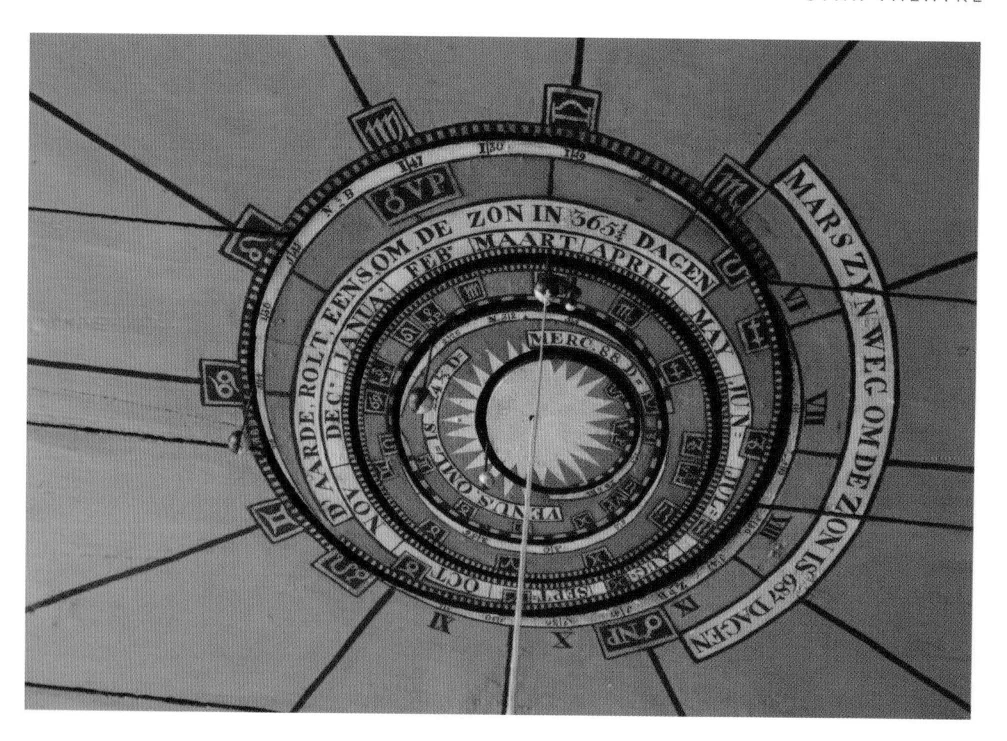

The ceiling of the Eisinga Planetarium, c. 1780.

the path of the Sun. This type of conjecture reappears repeatedly throughout the history of astronomy, for instance in the predictions of cosmic catastrophe by the Russian Immanuel Velikovsky in the 1950s, as well as today, with sensational reports in the popular press that the Earth is about to be hit by a giant asteroid.

Threats of cosmic extinction can, however, lead to architectural invention. Eisinga's reaction to such fears was to take the unusual step of constructing on his living room ceiling a working orrery showing the paths of the Sun, Earth, Moon and planets. This remarkable machine took seven years to build and was thus only completed for action some years after the predicted collision had failed to occur. Just as he finished it, Uranus, having been previously considered a star, was recognized as a planet by Sir William Herschel, but the Eisinga living room was not large enough to include its orbit – a not

Internal space of the Eisinga Planetarium showing Eisinga's bed.

unusual problem, as knowledge of the universe continues to expand and constantly exceeds the size of the model.

Eisinga's orrery is remarkable for several reasons. It is placed in an ordinary domestic room, rather than a palace or scientific institution, almost as a piece of ceiling decoration, with the room painted sky-blue. Eisinga's wife Pietje, with whom he had three children, wisely insisted that the room should continue to function as a living room in spite of the presence on the ceiling of the orrery, and that the cupboards under the dials showing the periods of the Moon and other astronomical phenomena should continue to be accessible for the storage of clothing, crockery and kitchen implements. Eisinga and his family quietly proceeded with their domestic routines while the planets rotated by above their heads. At the far end of the room was a small enclosure containing a bed, allowing the married Eisingas to sleep under the motions of the celestial bodies. It was powered by a set of complex machinery, all constructed by Eisinga out of timber wheels, leather bands, cogs with handmade teeth of bent nails, and large weights like those of a grandfather clock, to keep the solar system in motion.

Eisinga gained the experience needed to create this machinery from his experience of wool carding mechanisms, and indeed the effect is of a loom of the heavens. This machinery filled the attic of the Eisinga family house, located just above the living room, and was thus concealed from view. According to Adrie Warmenhoven, the current curator of the Eisinga house – now a museum – in his short book on Eisinga,

> When almost everything was ready, the pendulum was installed in the correct location, precisely over the cupboard bed. The pendulum however was too long to fit above the floor in the attic. To retain this pendulum would mean having to cut a slit in the floor to allow it to swing above the area of the bed. His wife did not like this idea, as she was very fond of her matrimonial bed, especially since her husband had already transformed their entire living room.

Eisinga could have shortened the pendulum, but it would have swung too fast and the planets moved at the wrong speed, so he had to recalculate many of the gears in the machinery. So do family affairs control the orbits of the planets.

Eisinga's planets moved at their equivalents' actual speeds and were therefore always in the correct position, the miniature globes in the living room corresponding to their rather larger counterparts in the sky. The orbit periods are written up in golden letters on the ceiling – Earth, 365.25 days; Mercury, 224.66; Mars, 687. Anyone entering the living room would be able to tell immediately the current positions of the planets – and thus be reasonably reassured that no apocalyptic collisions were impending. Eisinga's planetarium brings the model of the heavens – relating to the vast dimensions of the solar system and the movement of the planets at a scale of 1 millimetre to 1 million kilometres – into a typical Dutch home, above the everyday domestic routine of the inhabitants. The viewer is not yet quite within the paths of the planets, but the experience is different from that of simply looking at an orrery or armillary sphere; the device has become part of the room. The machinery in the attic echoes the medieval notions of a divine cosmic machinery that drives the solar system, but the prime mover, the being who makes all the other parts turn, is now made of wooden cogs and confined to the attic, a half-forgotten, hidden space within the house.

Eisinga appears to have had no further great ambition after building his orrery; he lived until 1828, occasionally lecturing on astronomy at the local university. Today Eisinga's orrery is still in position on his living room ceiling and continues to function. A portrait of Eisinga hangs nearby, showing him seated at a table, dressed in sober black clothing. The inscription on this portrait is taken from Horace, and reads: 'Arces Attigit Igneas.' He reached the fiery heights.

The Holy Theatre II

The theme of the holy reappears at various intervals and in various cultures. The dome of heaven, for Khosrow a sign of personal power, became an Islamic religious motif that, often based on the swirling firmament, featured in many mosques. The extraordinary ceiling of the Hall of the Abencerrajes in the Alhambra, Granada, has a star-shaped opening onto a dazzling 'sky' of cream-coloured decorative forms. Looking up into this dome is an almost trance-inducing experience, akin to an early piece of Op art; the visitor feels drawn upwards into this soft and geometric heaven, where depth and distance are so uncertain. The Hall of the Abencerrajes deliberately induces in spectators a mystical state. A number of Christian churches followed this Islamic lead to offer the dome as an internal heaven, such as Granada Cathedral, with its wonderful golden stars against a blue-green sky, and the easterly dome of St Mark's Basilica in Venice, where Christ emerges from a deep blue sky with silvery stars. These mosques and churches act as star theatres in the sense of providing a celestial background to the performance of a religious ceremony: the sky as location for spiritual beliefs. The intention in all these buildings is spiritual rather than astronomical; for the most part, the stars are shown as a decorative background, rather than as a scientific attempt to convey the position of constellations.

The Spherical Theatre II

'Heaven', wrote Henry David Thoreau in *Walden* (1854), 'is under our feet as well as over our heads.' Thoreau's paradoxical, almost Taoist thought could be interpreted in various ways. He implies, however, that Heaven – and maybe also the heavens – is rather closer to home than we might suspect, and to be found in an unexpected location. The hemispherical dome offering the illusion of being under the stars still leaves the viewer standing on a flat floor; the heavens are still above the viewer's head. If the dome becomes a sphere, and if the sphere rotates, then another kind of illusion is created, completely surrounding the viewer. The idea of the private star theatre for the sovereign, rather than a religious

Gottorf Globe, 1654–64, relocated to St Petersburg.

assembly, returns. The Gottorf Globe, a 3-metre diameter sphere, was built during the years 1654–64 specially for Frederick III, Duke of Holstein-Gottorp, at Schloss Gottorf, near Schleswig, north Germany – another man with a desire for his own, self-contained night sky. The globe was designed by the court librarian, adventurer and linguist Adam Olearius and built by the gunsmith Andreas Bösch. Olearius had travelled extensively in Persia and the Levant, and had heard in Persia of a glass globe with stars on the surface, in which one person might sit – a possible inspiration for his own, much larger, globe. The Gottorf Globe was constructed of metal strips lined with timber, and covered on the inside and out with canvas. On the outside surface was painted a map of the world, as understood at the time, before the use of lines of latitude. On the inside were painted remarkable images of the astrological signs and the figures for the constellations, with their stars set out with gilded polished nail heads, creating a baroque world that mixed mythical figures and stars.

The globe was open only to the friends and guests of the duke, who would himself trigger the lever to start the rotation, in the tradition of King Khosrow sitting under his rotating dome to display his power. The learned writer Eberhard Happel visited the globe and reported:

> Inside is a round table with a circular bench, around the axis, and I have myself observed that 11 people can sit inside the globe. Within are drawn with precision all the stars and figures of the skies, and also the circles of the heavens, so that it is greatly satisfying for a curious person to sit at this table, and see all the stars illustrated in their movements, the turning powered

Gottorf Globe with spectators.

by a water system flowing down the hill outside, corresponding to the movements of the actual skies.

Giovanni de' Vecchi and assistants, detail of the astronomical ceiling with constellation figures, Villa Farnese, Caprarola, 1575.

The shaft was therefore geared so that it could move in real time, showing the pattern of the stars to be seen in the Gottorf night sky, or it could be speeded up, anticipating the acceleration of time of later planetariums. A timber model of the Earth was fixed to the shaft, and rotated with it. No planets were shown. Within the interior, candles provided illumination, so that the duke and his guests, sometimes fortified by a helpful glass of wine, could sit inside and suppose themselves to be in their own miniature, lavishly pictorial version of the night sky.

The Gottorf Globe was placed in the main hall of a specially constructed villa; in the adjoining rooms were a cabinet of curiosities and a library of scientific volumes, with some rooms upstairs acting as living quarters. It is a playful object, but its implications are more complex than first appear. The spectator enters the sphere of the world to experience inside what is actually the exterior – the revolving night sky. They sit at the centre and see another, smaller version of the Earth, around which moves the Sun – though already it was well known that the Sun, not the Earth, was at the centre. The stars are in approximately the correct positions, but they are overwhelmed by the extraordinary painted figures of the constellations and the astrological signs, derived from earlier baroque ceilings such as the wonderful ceiling in the Villa Farnese in Caprarola. The viewer is surrounded by a spinning set of brightly coloured symbols – a crab, bears greater and lesser, bulls, water-carriers, nymphs, a ship, a swan, naked twins, Canis Major and Sirius, Orion forever pursuing the hare – all ordered by their position in the night sky. Here is being celebrated an old, already vanishing version of the cosmos as a set of pictorial symbols, interpreting sets of stars as terrestrial creatures and objects, which at the time had long been replaced by a more scientific explanation of the stars as objects with physical mass, moving at considerable speeds at great distances from the Sun. This vision of the figures of the zodiac still has an influence today, with the figures projected within the shows of many planetariums. The newest versions of the universe still rely on the charming figures shown in the Gottorf Globe.

The later history of the Gottorf Globe involves extended travels. In 1713 Duke Charles Frederick took the wrong side in one of the many Northern Wars, and lost his territory. The globe was given by the Danish king Frederick IV to a greater sovereign, Tsar Peter the Great of Russia. For over four years it travelled by ship and sledge to St Petersburg, where it was placed in the elephant house – the elephant sadly having died – and then within the tower of the Kunstkamera, in the centre of town beside the river Neva, where the tsar was assembling a collection of scientific objects. In 1747 the globe was largely destroyed in a fire, but reconstructed. The restless sphere continued to wander. In 1901 it was put on display in the palace at Tsarskoye Selo, on the outskirts of St Petersburg, the location also of the famous Amber Room, another extraordinary interior space, originally constructed in Berlin. During the Second World War Siege of Leningrad the globe was seized by the German Army and transported on a specially constructed railway wagon back to Germany, where it was put on display in a deserted hospital near Lübeck, not far from its original location in Gottorf. At the end of the war, now badly damaged with bullet holes, the globe was repatriated via Murmansk to Leningrad. Photographs from 1948 show the globe on a truck, and then being raised up on a rather unstable looking system of pulleys to the top of the tower of the Kunstkamera, the wall of which has been partly demolished to get it in.

The sight of this planet being restored to its position in the skies, at the end of a period of global catastrophe, seems a portent of the return of that other, much smaller planet, also hauled up precariously to the roof of the London Planetarium. The Gottorf Globe, the earliest prototype of a baroque planetarium, now rather over-restored, remains today in its position in the Kunstkamera tower, up in the Baltic skies. As Olearius had written at the time of its construction:

> One can make round objects
> Very easily roll.
> Nature has all things
> Set in circular rings.
> Thus nothing remains in place
> It must always move along.

The Spherical Theatre III

The Gottorf Globe led to other, less aristocratic, astronomical spheres. The aim was often as much philosophical as scientific. 'I am searching', wrote the mathematician Erhard Weigel, 'for the stillness at the heart of movement.' In 1661 Weigel build a 5.4 metre diameter globe of iron on the roof of his house in Jena, Germany – a curious precedent for the first planetarium ever to be built, on a Jena rooftop two and half centuries later. This globe revolved, had a fixed meridian, was painted with the zodiac on the outer surface, and was pierced with holes so that light from the outside would imitate the pattern of the stars to an internal viewer. Model planets could be attached to the surface via magnets, and moved along their paths – a step forward from the Gottorf version.

Weigel's sphere was not decorated with a set of baroque symbols, but for the first time attempted to create with its pierced holes the effect of starlight, with the light coming from outside the shell. Did he find stillness? Standing on a platform at the centre of his globe, he alone remained immobile while the stars span around.

The globe was demolished in 1692, but Weigel also built *Sternschränken*, or star cupboards – darkened shafts in which up to one hundred people could enter and look up to see the actual stars – and in 1670 what he termed a Globus Pancosmus, which was also enterable and equipped not only with holes for the stars but with fiery meteors and volcanoes, and now showers of rain and hail, accompanied by thunder, using technology which remains unexplained but was essentially a throwback to the dome of King Khosrow. In 1765 Roger Long, an astronomer interested in the measurement of distances to the planets and the stars and an inventor of eccentric musical instruments, created a similar globe at Pembroke Hall, Cambridge, which he termed The Uranium to celebrate the discovery of the planet Uranus several years earlier. This globe was rotated by a winch and supposedly capable of holding thirty people, though again they would have been rather pressed together in such a small space.

These globes provided a space for a limited number of spectators, squeezed in around the table. The actual dimensions remain comparatively small, but once again they suggest that the spectators had the impression of being out under the vastness of the night sky.

Baroque Theatre with an Orientalist Theme

The influence of these constructed spheres extended 'both' into fiction. The writer and philosopher Voltaire's short and absurd philosophical novel *The Princess of Babylon* (1768) relates the tale of a Middle Eastern princess who travels through the world and has various encounters with European culture, allowing Voltaire to satirize courtly life at home and abroad. The idea of the dome full of stars makes its return in this fairy-tale version of the East. Voltaire's domed room is similar to the Gottorf Globe, but scaled up considerably in size:

> an oval saloon, three hundred feet in diameter, whose azure roof, sown with golden stars, represented all the constellations with the planets, each in its correct position; and this ceiling, together with the dome, was turned around by machines as invisible as those which direct the celestial motions. A hundred thousand torches, enclosed in crystal cylinders, illuminated the exterior and interior of this dining hall. A buffet with stages contained twenty thousand vases and dishes, and opposite the buffet, on other stages, were a great number of musicians.

Étienne-Louis Boullée, Cenotaph for Isaac Newton, exterior, 1784.

The Voltaire planetarium is a Middle Eastern pleasure dome, mixing astronomy, dining and music.

The Spherical Theatre IV

In all these ingenious devices, a certain grandeur is lacking. The night sky has its own scale, its sense of immensity that should not be reduced to plaything or party space. Surely the universe of ever-expanding space requires something more serious? 'O Newton!' wrote Étienne-Louis Boullée of his proposal for an architecture to match the scale of the ideas of the scientist, 'I have conceived the idea of surrounding you with your discovery. That is as it were to surround you with yourself.' In 1784 Boullée designed a cenotaph for Sir Isaac Newton, the physicist credited with discovering the laws of gravity and of light. A vast sphere, 150 metres in diameter, in which would be located Newton's sarcophagus, sits on a massive

Étienne-Louis Boullée, Cenotaph for Isaac Newton, cross section with night sky.

base, around which grow two circles of cedar trees. The cenotaph would function by day and at night as two types of planetarium. The upper half of the sphere is pierced with small holes, through which sunlight could penetrate, giving the impression of a starry sky. In a curious inversion of reality, the stars would shine only during daylight, while during the hours of external darkness, artificial light would shine from a large orrery located at the centre of the building sphere. This orrery was to be composed of a physical representation of the planets, able to move mechanically along their correct celestial paths, the central light beaming the shadows of the planets onto the interior of the sphere. The night-time effect would therefore be similar to that of a planetarium, with the orrery assuming the position later taken up by the Zeiss projector. Access for human beings, reduced to the scale of insects by the vast size of the interior, was by a tunnel through the mass of the base, like a passage through the solid volume of a pyramid, leading into the hollow sphere, where they could stand, gazing up in awe at this extraordinary space. Boullée continued his description:

> The onlooker finds himself as if by magic floating in the air, borne in the wake of images in the immensity of space . . . The lighting of this monument, which should resemble that on a clear night, is provided by the planets and stars that decorate the vault of the sky . . . The daylight outside filters through these apertures into the gloom of the interior and outlines all the objects in the vault with bright, sparkling light. This form of lighting the monument is a perfect reproduction and the effect of the stars could not be more brilliant.

A similar architectural proposal, again with its external surface pierced with holes, was produced by Boullée's compatriot Jean-Jacques Lequeu in 1794, at the height of the French Revolution. Lequeu was a more eccentric figure than Boullée; he depicted himself in a self-portrait as a transvestite nun, and his work is often filled with barely concealed erotica, leading to the interest in his work by certain Surrealists such as Marcel Duchamp. His Temple de la Terre (Temple of the Earth), or Temple de la Sagesse Suprême (Supreme Wisdom), was produced at a time when Maximilien Robespierre's Cult of the Supreme Being had briefly replaced the Catholic Church as the state religion, with great public celebrations devoted to wisdom. Lequeu's sphere was to be built, as the accompanying text stipulates, in 'a pleasant fertile landscape, where the lands produce an abundance of all sorts of grains, fine pastures and rare plants'. The design exists only as a watercolour, in delicate tones, showing the section and front elevation. A globe of white marble, decorated on the outside with the continents of the Earth, is supported by a set of columns placed in a circle on a plinth. The interior is illuminated once again by

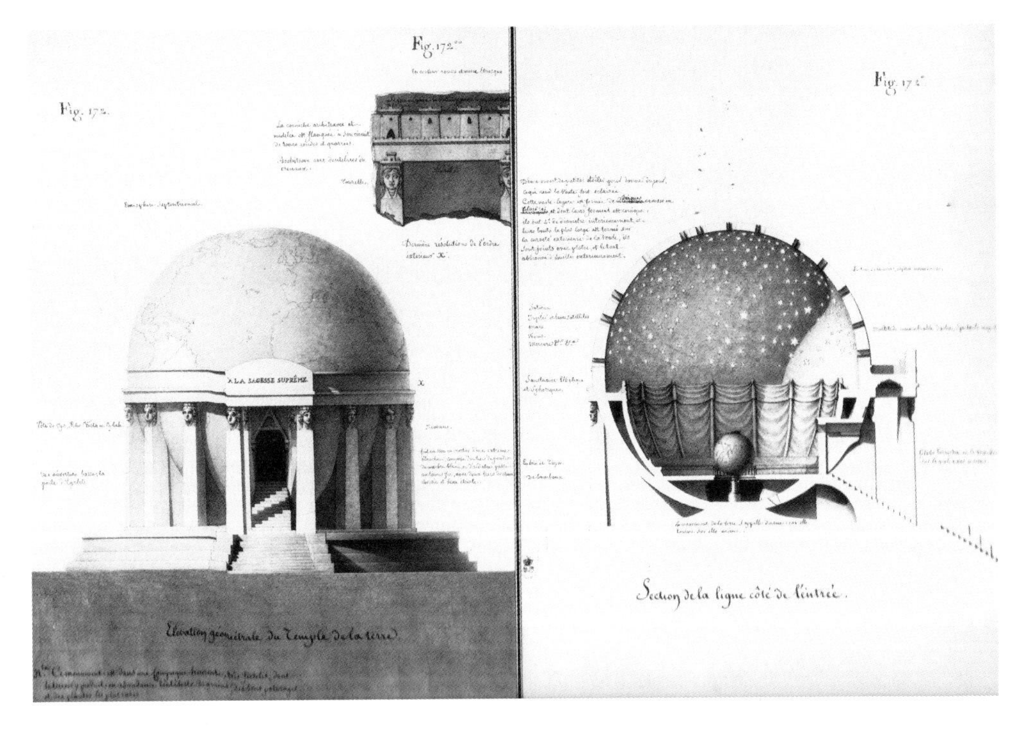

Jean-Jacques Lequeu, Temple de la Terre, c. 1794.

holes in the surface of the sphere, and a smaller terrestrial globe, apparently able to rotate, is placed at low level on a timber floor.

Lequeu was never a believer in the Cult of Reason – he was always wilfully irrational – and so his hollow sphere is a play on the notion of a rational universe. In contrast to the magnificence of Boullée's celestial sphere, here we are in an uneasy, disturbing depiction of the universe, a *mise en abyme* of descending spheres. Sitting on the actual Earth under the actual sky is the terrestrial globe of the temple, within which is found an illuminated sky, containing another terrestrial globe, which might in turn contain another sky. His Temple of Supreme Wisdom appears to be another evolutionary cul-de-sac, but its form anticipates the early German planetariums of the late 1920s, being part temple, part scientific device. The combination of the deification of science at a time of political violence seems familiar. La Sagesse Suprême, the final revelation of the wisdom of the universe, still awaits.

The Royal Theatre III – with a Musical Theme

The spheres of Boullée and Lequeu are theatrical in that they create an extraordinary astronomical stage set within which a performance of flickering starlight takes place. The showmanship of these great French globes filtered through, finally, into the actual theatre.

The orchestra plays enchanted music. The Queen of Night appears to plead with Tamino to rescue her daughter. The wonderful melody begins. The night sky as a background of sparkling stars appears in the great theatre set designed by Karl Friedrich Schinkel for a Berlin performance of Mozart's *The Magic Flute* in 1816 – a true star theatre. Schinkel's stage design has survived in a

Karl Friedrich Schinkel stage set for the Queen of Night in the opera *The Magic Flute*, illustrated by Karl Friedrich Thiele, 1815–16.

luminous aquarelle by the artist Karl Friedrich Thiele. Regular rows of stars rise from the clouds to form a dramatic backdrop above the lunar chariot of the Queen of Night, inducing a three-dimensional effect – a *trompe l'oeil* domed interior – to accompany Mozart's rapturous music. Looking in from the other side of the proscenium, the spectator remains just outside the realm of the Queen of Night, still in their own world but illuminated by the light of the stars.

The Spherical – and Rough – Theatre

With the Queen of Night we are already within a magical vision of the night sky, far from the considerations of science. Why anyway should a large-scale model conform to the conventional view of the solar system, with the surface of the Earth as convex? To display the surface of the Earth on the interior surface would be to completely invert the principles of Boullée and Lequeu, unintentionally leading to the bizarre theory of a hollow Earth, thus potentially upsetting all previous explanations of the nature of space. A globe 19.3 metres in diameter and made of timber was erected by the entrepreneur and mapmaker James Wyld in London's Leicester Square in 1851. The inner surface showed the Earth as a concave form, in relief, with thousands of plaster-cast models of geographical features, including erupting volcanoes with flares of red cotton wool, and icy mountains of sparkling crystal. Wyld's globe attracted large numbers of visitors: 1.2 million in the two-year span of its existence. The public could ascend via a timber staircase and from platforms view the scene, lit from above by a large opening in the crown of the sphere. The lower part of the exterior of the globe, which was surrounded by an external corridor, was decorated with star maps, painted by a scenery designer.

Wyld's intention was that visitors should have an impression of the Earth as a whole, but his globe also has astronomical implications.

James Wyld, Great Globe, 1851.

It could be considered as an anticipation of the highly eccentric theory of the concave, hollow Earth, in which the surface of the Earth is considered to be the interior rather than exterior of a sphere. This theory was first proposed by Cyrus R. Teed in 1886, well after the construction of Wyld's globe, but the two interlink. What would it imply for the night sky if it were inside rather outside the sphere of the Earth? Teed explained the obvious problem of the apparent locations of the Sun, Moon, planets and stars as mere illusions created by the curvature of light. The hollow Earth theory, which unsurprisingly has attracted few adherents, was later updated by the Egyptian mathematician Mostafa Abdelkader, who in 1981 proposed that space-time contracts as it approaches the centre of the hollow sphere, thus explaining how rockets are able to rise from the Earth's surface and allowing infinite space for extra galaxies and other distant features of more conventional astronomy. Just how this thesis might have been demonstrated in a theatre or planetarium remains a mystery.

Numerous other globes of vast size were proposed in the following decades, fulfilling an increasing desire to model the Earth and the solar system at an architectural scale, as though the version provided by the prime mover were no longer satisfactory and man could improve on the original. An unconstructed project in Chicago featured a globe that would hold 10,000 people placed on the shoulders of a 150-metre-high statue of the giant Atlas. The anarchist and writer Élisée Reclus proposed for the Exposition Universelle of 1900 in Paris a model of the Earth as a sphere 127

Élisée Reclus and Louis Bonnier, Globe Terrestre for the Paris Exposition Universelle of 1900.

metres in diameter, enclosed by a remarkable egg-shaped exterior skin containing a spiralling ramp. A smaller celestial globe at a diameter of 60 metres, named the Cosmorama was designed by the architect Paul Louis Albert Galeron, with the constellations on the exterior, to be viewed from a travelling cable car which would wind its way up and around, and with a smaller internal sphere into which spectators could enter to see a display of the movements of the planets. At this point the desire to construct vast models of the Earth and its Solar System seems to have been exhausted.

The Spherical – and Immediate – Theatre

Finally in this long sequence of theatre of the stars, at the beginning of the twentieth century, a much more modest astronomical globe was built in 1913 for the Chicago Academy of Sciences. A mere 5 metres in diameter and constructed of steel, it adopted the ideas of Weigel and Boullée, with 692 holes drilled in the surface to imitate the stars. The globe had enough internal space to contain ten squeezed-in spectators, who were moved into place via a mechanical platform. The sphere revolved slowly to give the impression of the turning of the night sky. An illustration showing a section through the sphere shows a girl standing with her head at the exact centre, an Alice in Wonderland figure staring up reverently at the artificial stars as wheels and cogs rotate the sphere in a device halfway between those astronomical machines and the projected light technology to come. The device is spherical, but also immediate, the effect is direct but mysterious, again the spectators are in a space both everyday and comparatively confined, and opened up to some much larger celestial illumination. The Atwood Sphere, which is today located within exhibition rooms of the Adler Planetarium, combined the mechanics of the Gottorf Globe with the effects of the Boullée sphere. It was the last of this great sequence of astronomical spheres constructed before the appearance of the actual planetarium.

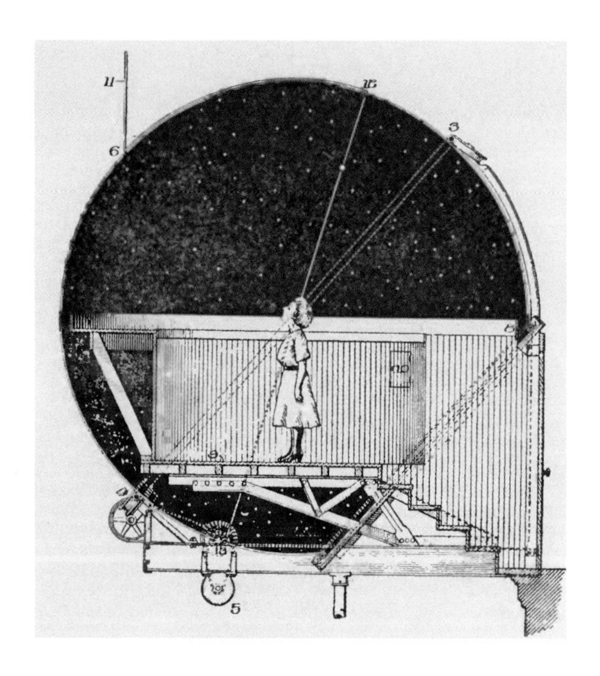

The Atwood Sphere, Chicago, 1913, *Popular Science*, vol. LXXXIV (January 1914).

EACH OF THE VARIOUS PREDECESSORS of the planetarium listed in this chapter creates its own version of the night sky. As the nineteenth century progressed, the wonders of nature began to have to compete, with increasing lack of success, with ingenious man-made simulacra, invented theatres influencing or even replacing the natural world. The spheres, domes and machines were part of a much larger attempt to create an imitation of the natural world, as increasing numbers of people moved into the cities and had little knowledge of real natural phenomena. By the end of the nineteenth century there were a wide variety of such artificial recreations, sometimes containing actual elements of nature in artificial surroundings: the zoo offered its real animals in a fabricated environment; the natural history museum had its theatrical scenes of forest plains and deserts filled with the appropriate stuffed animals; the aquarium set off real fish against artificial marine backgrounds. Panoramas, setting out a 360-degree painted reproduction of a city, a landscape or an event such as a battle, were constructed in large buildings, often with a dome painted to resemble the sky, in many European and American towns, and were highly popular in the nineteenth century.

Most importantly, the invention of cinema, based on the projection of light, recorded and reproduced the actual world as a series of flickering images on a screen. The cinema moved away from material recreations, constructions of timber and plaster, to the immaterial, the creation of illusion through light. The cinema would provide the projection technology that would be adapted to become the planetarium projector. But unlike cinema, which utilizes a recorded version of the world that never changes, the planetarium is slightly different at each performance. 'The cinema', wrote Peter Brook in *The Empty Space*,

> flashes on to a screen images from the past. As this is what the mind does to itself all through life, the cinema seems intimately real. Of course, it is nothing of the sort – it is a satisfying and enjoyable extension of the unreality of everyday perception. The theatre, on the other hand, always asserts itself in the present.

Planetariums, and some of the globes mentioned above, contain aspects of the cinema, with their use of the projection of light, but they have the quality of theatre as the spectators, seated around the great projector, watch a live performance, the movement of the planets always a little different, the voice of the lecturer adapting to the nature of the given audience. The planetarium, like any theatre performance, repeats, but also varies according to the mood in the dome.

As for that planetary show that appears each night and vanishes each morning, as though distantly recalling the myth of the Egyptian Sun goddess and the daily passage through the underworld: it is time at last for the true theatre of the stars. The luminous projection system at the centre of the room begins to stir of its own accord. The calm voice of the lecturer can be heard. A voice, emerging from some childhood memory, is suddenly recalled. The show on the rooftop of Jena is about to start. It is time to take our seats before the lights go down and the stars emerge.

TWO

PLANETARY PROJECTION

In 1924 a new construction appeared on one of the roofs of the Zeiss factory at Jena, a small city in eastern Germany. On a neighbouring roof there already stood the dome of the Zeiss observatory, the *Sternwarte*, within which Zeiss could test out telescopes for observing the night sky. All around on the skyline of Jena stood other towers, Gothic gables, domes, spires and chimneys, belonging to the churches, factories and warehouses of the city. The new construction was a second dome, larger than the observatory, composed of a complex network of thin, metal rods – the first tentative geodesic dome.

Unusually, it was constructed from the top down: first, the rods were attached together on what would become the crown, then this network of elements was raised up on cables strung from two timber scaffolds. The sections below were then added one by one, until a hemisphere 16 metres in diameter and made solely of interlinking rods stood independently on the roof. This dome was then covered with wire mesh and plastered with a thin layer of cement. Its interior was lined with linen fixed to a metal lattice. A curious machine, the first of its kind, about a metre and half in height and vaguely resembling a kind of scaled-up kitchen instrument, with assorted nozzles and drums, was brought up into the roof and placed on the floor in the centre of the dome. Mechanics in their workshop overalls tested out its moving parts, adjusting the cogs and tightening the belts, while a small man with a moustache, dressed in a dark work suit, directed operations. Another, larger man, bearded and well dressed, looked on and took notes. The background lights were lowered until the inside of the dome was completely dark; one after another, projectors within the machine were illuminated, cogs and belts began to turn as other moving projectors came into play, and circular lights flickered up across the interior linen lining. Gradually these blurs of illumination were brought into focus to form pinpoints of light.

The spectators would have recognized familiar patterns among these points of light, linking them up to form constellations with names vaguely remembered from ancient myths, others that slipped past at unfamiliar speeds, a gradually evolving artificial night sky, never seen before.

THIS CONSTRUCTION, ON THE roof of the Zeiss works in Jena, was the first planetarium, known as the *Sternentheater* or star theatre. It worked based on the projection of light and was designed by the Zeiss optical engineer Walther Bauersfeld for Oskar von Miller, the founder of the Deutsches Museum in Munich. The name star theatre is nearer to the construction's actual function than the word planetarium, for the apparatus within the dome showed mainly stars with the planets as an addition, and it had the qualities of a theatre – a show was put on for an audience. It was created under particular conditions, linked to the social, political and artistic circumstances of the Weimar Republic.

Oskar von Miller, sometimes known as 'Red Oskar' for his anti-nationalist leanings, was an idealist devoted to the use of new technologies for the benefit of the German people. Miller was also director of Bavaria's largest electricity company, AEG, and had been responsible for late nineteenth-century technical progress

The Zeiss dome under construction.

in building distribution systems for electrical power, thus bringing electricity to German towns. He saw electricity not just as a source of power but as a civilizing force. In 1903 Miller had the grand notion of establishing in Munich a national technical museum, where people could be shown historical and contemporary technology. Construction of the museum took over two decades, being interrupted by the First World War and the consequent economic collapse and shortage of funds for any such cultural projects.

Miller was keen to include a section on astronomy in his museum. He was advised by the German astronomer and director of the Heidelberg observatory, Max Wolf, who had spent many years cataloguing stars, tracing asteroids and studying through the developing technology of astrophotography the then little understood phenomenon of dark nebulae, the apparently black areas of the night sky which in fact contain a fine opaque dust. Wolf was interested in how increasingly complex ideas of astronomy could be conveyed to people with little background in science, and encouraged Miller to include a demonstration of the night sky in his museum. In 1913 Miller had the idea of building two planetariums for his museum. They would be called, somewhat confusingly, Copernican and Ptolemaic, in that the former would show the movements of the solar system to a moving observer, and the latter to a fixed audience, but both naturally with the Earth going around the Sun. Visitors would thus be shown how the planets moved around the Sun in a model, and also be given a theatrical show of what the night sky looked like. Model and theatrical show were considered at first as separate operations. These two planetariums were to be located in the tower of the Munich museum, just below a domed room containing a massive Goertz telescope, so that – and this was to be a continuing theme – directly observing the heavens would be linked to understanding how the heavenly bodies moved, and information derived from the telescope could be interpreted in the planetarium. Somewhere in the whole Miller enterprise there was a curious contradiction, which would continue to run through stories of the creation of simulations of natural phenomena: the glare of electrical city lighting, which Miller himself had brought to German cities, made the stars difficult for city-dwellers to see, so an artificial sky was now necessary as a replacement for the real experience. The natural world, no longer fully visible due to the advances of lighting technology, would be replaced by an ingenious simulation.

Miller's Copernican planetarium, constructed in 1913, was a fascinating but ultimately unsuccessful machine. Like in the earlier Eisinga orrery in the Netherlands, the planets were mounted on the ceiling with a complicated system of belts and wheels to drive them. Orreries had always had the drawback that the observer stood and watched the movements of the spheres from outside, detached and not part of the system. Globes visitors could enter, such as the Gottorf Globe, were attempts to remedy this, but they were never capable of demonstrating much detail. In the Munich machine, however, the Earth was represented by a small wagon, mechanically propelled along its orbit, with a spectator inside observing the other planets through a periscope. The observer had finally found a way to escape his detached position and to become an integral part of the celestial mechanism. This was an innovative proposition and one of the rare occasions when such a dynamic version of the solar system and its observer has been attempted. It could be seen as linking the notion of the planetarium to certain aspects of Einstein's theories of relativity – at the time

Copernican Room, built 1923(?), at the Deutsches Museum, Munich. Photo from the 1940s.

gradually becoming accepted by the scientific community – which question the position of a supposedly neutral observer within any system. In his early studies, before the entire theory passed beyond the easy understanding of the layman, Einstein made use of the word *Beobachter* (observer), and was fond of using the example of observers, equipped with clocks, in moving railway carriages to explain the relative nature of time. The machine in the Munich museum was not explicitly concerned with such theories, but by putting the observer in motion it introduced a form of relativity into what had previously been a simpler matter of a fixed observer. In the museum the observer in the carriage would be given a view of the solar system, with everything, including himself, being in motion and therefore offering a continually changing viewing position.

This mechanical planetarium had charm but it was too clumsy, too mundane, lacking any sense of cosmic illumination. Despite its high ambitions, this planetarium was rather basic. Photographs from the time show a portly Bavarian man dressed in knickerbockers standing in the carriage and staring through the periscope, as an officious museum curator with a long stick points out which planet to look at. The idea was better than the final construction: the general effect was of an individualistic machine, of the solar system reduced to a fairground ride. It was a version of those earlier attempts in nineteenth-century panoramas to get the observer to move within the scene in imitation ships and wagons, and a precursor of those in the World's Fairs of the 1950s and '60s, where the viewer was carried in a small train past theatrical sets showing life on the Moon or under the sea. Travel in the Munich carriage was fun, but only one person could observe the solar system at any one time, and the view through the periscope was limited – effectively little better than looking at a set of model spheres moving in and out of view. It was difficult to focus on more than one sphere, and they looked more like balls of various sizes than luminous planets. The periscope view was also curious, an anticipation of the view through a U-boat periscope at other ships moving on the ocean's surface, one of the defining viewpoints of the later First World War. This planetarium remained in place until the bombing of Munich in 1944, during which the museum was badly hit and partly destroyed. All that remains today of Miller's innovative Copernican machine are a collection of the spheres that once boldly represented the planets.

The second planetarium, conveying the actual experience of the night sky, would require something rather more inspired, an invention appropriate to the technology of the twentieth century. It was called Ptolemaic because it would show the movement of the planets to an observer in a fixed position, rather than one trundling along in a carriage, as in the first machine. It would of course show the Copernican system as a visual experience, combining science and entertainment. In 1913 Miller approached the directors of Zeiss Optik in Jena. Zeiss was a traditional family concern that had for decades designed and manufactured all kinds of optical instruments, including telescopes, cameras, medical machines, gun sighting devices and all kinds of lenses. Early ideas for the Ptolemaic planetarium concentrated on the idea of a large globe of stars that would revolve around the spectator, in the tradition of the Gottorf Globe and the Atwood Sphere in Chicago, but the first attempts were unsatisfactory – the proposed size of the globe made it hard to revolve, and the whole thing lacked celestial panache.

The enterprise was put on hold with the advent of the First World War, but at Zeiss, Miller had encountered a remarkable man, Walther Bauersfeld, who would take the project far beyond the limited confines of the original idea and who quickly proposed abandoning the tedious task of moving metal globes by instead using a projection system.

Walther Bauersfeld's rooftop dome, on the Zeiss factory in Jena, 1923.

Into a morose meeting of engineers, enter Bauersfeld, with enthusiasm:

> Why should we build such a complicated machine? I think we should project the images of the sun, moon and planets on a metal dome. All these mechanics can be replaced by a simple construction at the centre of the dome, which contains the projectors for the celestial bodies.

Almost before Bauersfeld had finished speaking, scientific director Professor Straubel, also at the meeting, said, 'And then we can also project the fixed stars onto the dome.'

Bauersfeld noted, in his spidery handwriting, in his notebook: 'At this moment the projection planetarium was born.'

Bauersfeld was a highly inventive optical, mechanical and structural engineer. His background at Zeiss had been in the various aspects of photography and cinema, in particular in the field of stereo-photography, which uses multiple photographs to provide an illusion of three-dimensional depth, and its application in combining aerial photography of a landscape with two-dimensional maps, as used during the war for aerial reconnaissance flights. Bauersfeld was a man with many skills and with an understanding of different kinds of projection systems, of complex, three-dimensional geometry, and also, eventually, of innovative building construction. All these skills would be necessary for the evolution of the first real planetarium.

Bauersfeld quickly proposed replacing the hefty mechanical system with a projector. He described his idea as follows:

Walther Bauersfeld projector sketch, c. 1920.

> The great dome will be set up, its white interior surface shall be used as the projection surface for a multitude of projection apparatuses arranged at the centre of the dome. The interconnected positions and movements of the smaller projectors will be linked by appropriate gears, so that images produced by the projectors on the dome will show our eyes the visible stars in their position and movement, just as we are accustomed to see them outside in nature.

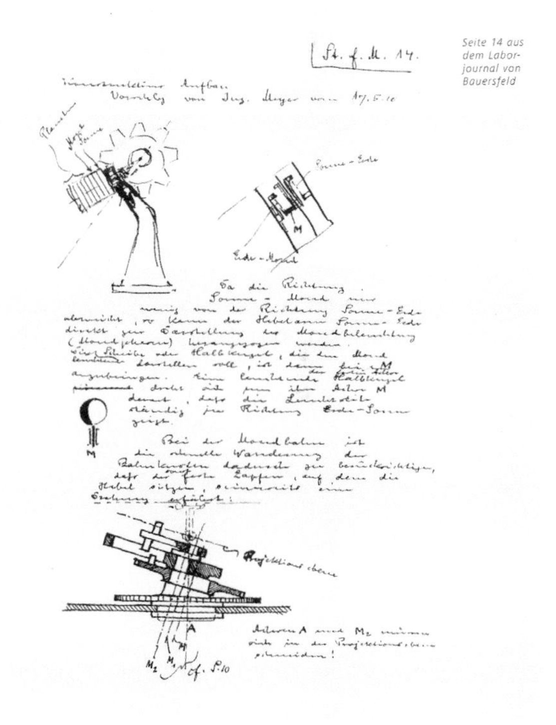

Within this dome both spectators and projector would remain fixed

Zeiss projector Mark I, 1923.

in place while the projections of the night sky moved around them. Bauersfeld's idea derives of course from the cinema – among his other activities, he had been involved in cinema projection systems – and from the whole notion of projecting light to create an illusion of reality, which had been familiar since the time of the

Lumière brothers thirty years before but was now applied not to adventures, romances and expressionist vampire movies, but to astronomy. The stars of the silver screen would be replaced by the actual stars.

The mechanical problems were now reduced to designing the projector; the previous efforts throughout the history of astronomy to create a complicated moving dome became suddenly redundant. However, while in the cinema there is only one projector and one beam of light shining onto a two-dimensional screen, to create an artificial night sky the planetarium would require many beams of light moving independently of each another, shining up onto a three-dimensional, hemispherical screen.

The Zeiss technicians, given the task of designing a projector that had to deal with fixed stars as well as planets on elliptical orbits moving at different speeds, all rotating around the Sun, declared the task impossible. Bauersfeld, so the story goes, reached the solution over a weekend. His elegant freehand sketches still survive in the Zeiss archive, and show how he patiently and with considerable imagination worked his way through the various problems of intricate three-dimensional geometry, delicate clock-like mechanisms and varying light strengths. He was inventing not just the mechanics of a projection system but a new way at looking at celestial bodies moving across the skies, and a new building typology. Over a period of five years he and his colleagues at Zeiss worked out the complex geometry, intricate mechanics and multiplicity of lenses needed to imitate the effects of the moving planets and the thousands of fixed stars.

The earliest Zeiss prototype projector, known as the Mark I, used on the roof at Jena, elegantly solved the problem of how to produce both a background of fixed stars and moving lights for the planets. The fixed stars, which for the purposes of the show were presumed not to move in relationship to each other, even though of course up in the heavens they are actually moving at considerable speed, were beamed from a sphere within which was fixed a series of projectors, each shining onto a particular area of the planetarium dome. There were six classes of brightness for these 4,500 stars, so that some would appear further away than others. Also within the globe were projectors with soft light, for the Milky Way, and others shining out the names of constellations and the zodiac. Just below this globe, on a metal rod set at 23 degrees to the horizontal, representing the tilt of the Earth, were mounted further projectors capable of independent movement, for the Sun, the Moon and the five planets Mercury, Venus, Mars, Jupiter and Saturn (Uranus and Neptune being for the moment omitted). The whole machine could swivel to imitate the turning of the Earth, and the planets could be run at various speeds to create the passing of a year in fifty seconds, two minutes or four minutes. All this required delicate and highly complex systems of motors and cogs, reminiscent of some of those earlier orreries but now applied to a true celestial modulator.

The projector was installed briefly in the Deutsches Museum in Munich for a trial, and then placed in the temporary rooftop dome specially designed for the purpose by Bauersfeld. He was assisted in the design of this dome by the engineer and expert in thin-shell constructions Franz Dischinger, who supplied the expertise in construction techniques that Bauersfeld lacked. The dome was extraordinary; it was the first-ever geodesic structure, based on a pattern of triangulated icosahedrons, similar to that which Bauersfeld had used to establish the relationship between the beams of the projector and the abstract geometry of the dome. The geometry of icosahedrons developed by Bauersfeld to work out the pattern of projectors for the fixed stars, suitably amended, was used to construct a network of metal bars, forming the structure of the dome. The material structure of the architecture followed therefore from the immaterial pattern on the dome set out for the beams of light.

The geometry for the light beams Bauersfeld already knew, but from where did the idea of using icosahedrons as a material structure come? One hypothesis suggests that the origins of the use of a geodesic structure for a dome lie in natural history. Among the buildings owned by Zeiss in Jena was the Villa Medusa, formerly the home of the great nineteenth-century naturalist Ernst Haeckel, who had studied the morphology of natural forms and produced wonderful illustrated books showing the details of microorganisms. The villa was named in reference to his favourite creature, the octopus. Among Haeckel's drawings are several of radiolaria, minute aquatic organisms whose bodies have a structure of icosahedrons. Haeckel had models made, by the father-and-son glass artists Leopold and Rudolf Blaschka, showing the structure of the radiolaria. The models and drawings of the radiolaria and the dome on the Zeiss roof are remarkably similar. Bauersfeld must have been aware of Haeckel's work and, consciously or unconsciously, used the structure to evolve his lightweight dome. The minute radiolaria emerging from the darkness of the undersea realm may have been the prototype for a building devoted to the skies.

The geodesic dome would of course develop its own history. The Bauersfeld dome was evolved in the 1950s and '60s by the American architect R. Buckminster Fuller, but without giving any credit to Bauersfeld, to create a new generation of domes of sizes ranging from a few metres to a proposal to cover part of Manhattan. Images of the Bauersfeld dome had actually been widely circulated in the U.S., for instance in books by the Bauhaus artist László Moholy-Nagy. Publications on Fuller occasionally show the Haeckel images of radiolaria, again without credit to Bauersfeld. Fuller's domes differ from those of Bauersfeld in that the geodesic structure is usually exposed and filled in with panels, whereas Bauersfeld concealed his structure of rods within the concrete skin. Bauersfeld was not particularly concerned with external architecture; his concern was with creating his own version of a much greater exterior space within the dome of the planetarium.

The planetarium on the roof, which soon became known as The Wonder of Jena, attracted 50,000 visitors from all over Germany in the first weeks of its opening. Long queues formed along the rooftop. The delights of the star theatre were clearly highly attractive to a population emerging from national military defeat and loss of lands in its east, and at a time of economic and political instability in the Weimar Republic. If everyday life was uncertain, the illuminated display of the solar system offered both scientific reliability and entertainment. Who could not be inspired by the sight of the night sky gradually emerging in the darkness of the dome and of each of the planets following their given path? The artificial night sky was both dependable and magnificent; in addition, for those who quite reasonably until then had no idea of, and perhaps little interest in, the principles of astronomy, the motions of the celestial bodies were now perfectly understandable.

The rooftop planetarium in Jena was only a temporary construction, and removed after a few months. The projector was installed in a room at the Deutsches Museum occupying a dome 10 metres in diameter, thus finally supplying Oskar von Miller with his second planetarium. In its new position the planetarium was again popular, attracting large numbers of visitors. The elegance of the Bauersfeld system compared to the clumsiness of the Copernican planetarium in the next room was obvious; Victorian mechanics were no longer enough to convey the delicate movements and subtle luminescence of the solar system.

Visitors to the planetarium enthused over the spectacular nature of the display, almost as an improvement on the actual night sky. The constructivist painter and graphic designer Walter Dexel wrote:

> Joyfully and majestically the heavens move over us, clear and clean, as is rarely to be seen in reality. Sun, moon and planets move along their paths, the fixed stars shine, the Milky Way shimmers in the gleam of its countless stars – an impression that competes with reality . . . We almost believe ourselves to be in the open air.

The considerable interest aroused all over Germany by the star theatres on the Jena roof and in the Munich museum suggested the need for a series of permanent buildings. The city government of Jena proposed to build a planetarium in Jena's Prinzessinnengarten, with a larger dome, this time 25 metres across and featuring the first Zeiss Mark II projector. Since this planetarium in the park was a more official construction compared to Bauersfeld's dome on the factory roof, the responsibility for its design passed from engineers to the Jena architects Schreiter & Schlag, who produced a domed classical pavilion with a formal entrance porch and a row

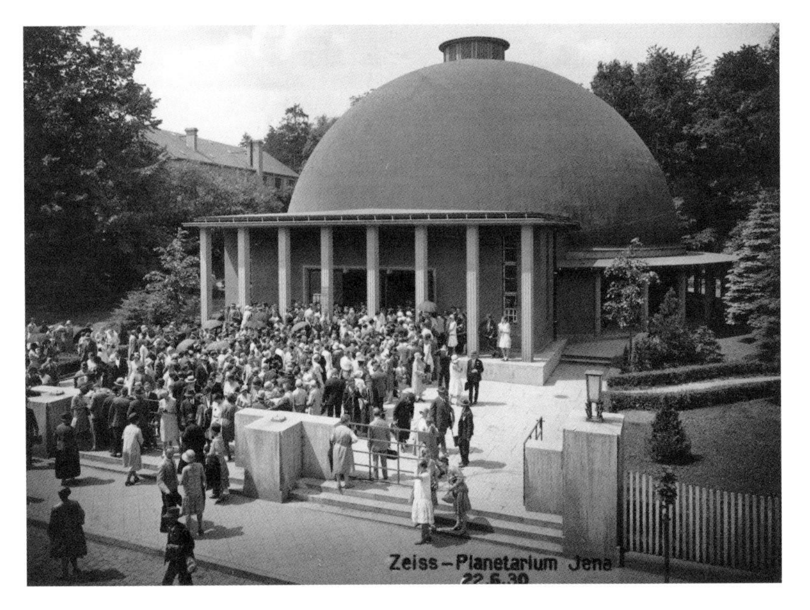

Crowds of people outside the Jena Planetarium in the Prinzessinnengarten, 1926.

Model of Adolf Meyer's project for the Jena Planetarium, 1926.

of columns around the exterior at ground level. The dome was once again designed by Bauersfeld, as another geodesic system that refined his original version to produce a shell of extraordinary delicateness, its thickness so reduced as to be thinner in proportion to the interior space as the shell of an egg is to its own interior. The minimal elegance of Bauersfeld's dome rather contrasts with the half-hearted gesture of the stripped-down classicism of the porch.

The design of the Jena planetarium aroused considerable discussion as to what architectural form a planetarium – a completely new building type – should take. The old medieval question of what lies beyond the last visible sphere returned in a new form. What kind of exterior would be suitable for a building that contained the heavens? What could be external to an internal space that was both on the everyday scale of the spectators and also of the universe? And was the planetarium a cinema, a civic building, a temple, a panorama, a theatre? These questions would continue to be raised throughout the design of later planetariums, and remain unresolved today, beyond the general acceptance that there is no particular answer. A planetarium is usually domed but in essence can have whatever exterior form the architect proposes, so long as it can contain the interior space of the hemispherical screen and projector.

A rival modernist design without decoration, with a simple slightly parabolic dome, raised up to first-floor level and with an entrance area at ground level, had been proposed by the Bauhaus architect Adolf Meyer. Meyer compared the purity of the construction system of the Bauersfeld dome to the decorated forms of traditional architecture: 'The image of the illumination of the Zeiss dome is completely different, here form and construction reveal

Zeiss projector Mark II, 1920s.

themselves in their pure mass, the construction system shown by the outer dome is an image of crystal clarity and unambiguous form.' Meyer deliberately used terms such as light, projection and crystal – which actually belong to the interior Zeiss projection system – to describe his architecture. Meyer's proposal remained unbuilt, but its simple lines and parabolic dome would a few years later influence the constructivist planetarium in Moscow.

Bauersfeld and the Swiss engineer and astronomer Walter Villiger had already come up with the Zeiss Mark II projector, the definitive model that would, with numerous improvements, last several decades and be supplied to planetariums all over the world. This remarkable machine could now project the positions of the planets and stars as seen from any latitude and in any year in the past and future. It took the form of a dumbbell, with two sets of

Barmen Wuppertal Planetarium, 1926.

star globes and planetary projectors, one each for the northern and southern hemispheres. Special projectors now provided the Andromeda Galaxy and Orion Nebula; the Sun projector was capable of varying tones of light; and the phases of the Moon and eclipses could be portrayed. The immaculate imitation of the delicate light effects provided by the night sky included small but swiftly moving projectors to create the paths of comets and meteors. This mighty beast, almost as impressive in its physical form as the lightshow it was capable of producing, was mounted on a frame of lattice beams, in turn placed on a wheeled carriage so that it could be moved to one side if the hall was to be put to other uses. Besides the magnificence of the Zeiss projector, which sat in the centre of the dome machinery and dominated the room, all other equipment was rather more basic. The lecturer, whose role was to be part priest, part schoolteacher, part actor, was a vital aspect of the planetarium show, since he provided a human presence in what was otherwise a mechanical performance. He stood at a wooden pulpit, controlling the individual lights of the projectors, speeding up and slowing down the show as necessary and providing a commentary on the cosmic phenomena the audience was watching. In the early shows, before the introduction of accompanying music, the lecturer's voice was the only sound to be heard in the hall, so the tone of his voice and his manner of speaking were vital for the atmosphere within the room. The audience sat on traditional wooden chairs arranged in a circle around the projector, and had to crane their heads back to see the show. The interior of the planetarium was clearly a distinctive kind of space, with its own identity, producing its own highly individual show.

Planetarium fever gripped Germany. The planetarium became a must-have for any respectable city of the Weimar period. Zeiss put into production 25 of the dumbbell Mark II projectors, which were all quickly snapped up. Eleven new planetariums were

Wilhelm Kreis, Düsseldorf Planetarium, 1926.

constructed in Germany in the brief period from 1926–30. The first to open was actually in Barmen, constructed at great speed to beat the opening date of the building in the Jena Prinzessinnengarten. The Barmen planetarium again resembled a classical temple, on a base of brick and stone, but with the standard Bauersfeld dome. It was located in the central city park and surrounded by trees. Approached up a long flight of steps, the entrance was flanked by two statues, and a small stone globe was placed at the summit of the dome, the whole giving the effect of a faintly disturbing miniature temple dedicated to some unknown deity. Leipzig, Düsseldorf, Dresden, Berlin, Stuttgart, Hamburg, Hannover and Mannheim soon followed, each planetarium varying greatly in appearance.

None of these early German planetariums is dull and almost all show considerable invention in considering how the dome would be integrated into the city. Dresden gained a *Neue Sachlichkeit* (New Objectivity) styled temple, designed by the city architect Paul Wolf, an architect whose career would later flicker apparently without problem from traditionalism to modernism, fascist monumentality and socialist mass urban planning. His first design resembled Florence Cathedral with a Brunelleschi-like dome, but the actual Dresden building was more modest and delicate, and was crowned by a small planet on the roof, an ancestor of the later missing planet in London. The city architects of Berlin produced another small temple, beside the Zoological Garden, surrounded by statues of astrological figures. More ambitiously, Hamburg created a turret at the top of a circular, brick Hansa-style tower built in 1916 and formerly used for water storage, located in the main urban park, with the stars and planets of the night sky replacing the previous volume of water. It was necessary to take an elevator up to the planetarium, beside which there was a viewing platform so that visitors could check the actual night sky against the artificial version within the turret. Hannover also chose the option of a dome in the skies, at the summit of a remarkable new Brick Expressionist building for the local newspaper the *Hannoverscher Anzeiger*, designed by Fritz Höger, architect of the brick Chilehaus (1924) in Hamburg and a specialist in buildings faced with clinker brick. The Hannover building, with its virtuoso brickwork and copper-clad dome, was the most prominently urban of all the new planetariums and demonstrated how the form of the dome might be integrated into the dynamic and stylish architecture of northern Germany.

Fritz Höger, Hannoverscher Anzeiger building, Hannover, 1927–8.

Zeiss Mark II Dumbell projector, 1920s publicity image.

Often the new planetariums were initiated by city authorities. Stuttgart, designed by the city architects, created another room in the sky, with a projection room in the upper storey of the Hindenburg building, opposite the main station, revealed only by large illuminated letters high up on the city skyline spelling out PLANETARIUM. At Nuremberg the planetarium, another building produced by the city architects, used a simple drum of brick, but with the projection space raised up to first-floor level to allow additional accommodation on the ground floor, a design move which was followed by many post-war planetariums. As part of the large-scale trade fair GeSoLei, a grand show typical of the time combining health, social welfare and physical exercise, Düsseldorf included a planetarium, a neoclassical Valhalla in concrete faced with clinker brick by the architect Wilhelm Kreis. At the time this was largest planetarium in the world, with a hall 30 metres in diameter – the maximum internal space the Zeiss projector could beam onto – and accommodating 2,500 spectators. Ingeniously, the interior hemispherical screen could be raised up to expose a balcony running around the interior, used to expand the seating when the space was used for musical performances. The ground floor, in an enterprising spirit of mix of science and sport, provided accommodation for a rowing academy, the rhythm of the oarsmen no doubt matched to the paths of the planets in the great hall above.

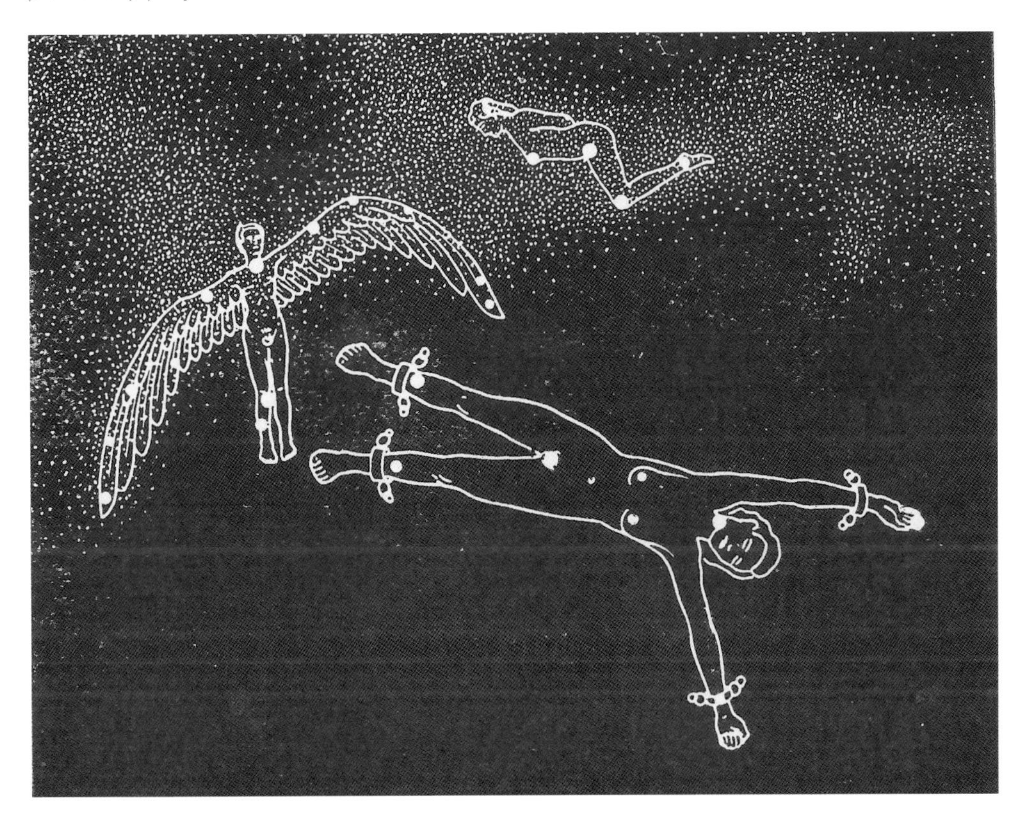

Zeiss star figure projection, 1920s.

These first planetariums were highly popular, the crowds flocked in to see the new mix of science and entertainment. The reputation of the Wonder of Jena spread rapidly, and visitors began to arrive from all over the world. Spectators young and old were equally enthralled by the novelty as much as the educative value of the performance. It became standard practice for German schoolchildren to be taken to see the star show.

But the planetariums also had to compete with the cinemas, which were able to alter their shows each week and offered a wide variety of types of film, while the planetarium tended to put on the same repeated display. The stars of the sky had to contend with the charms of vampires, murderers, deranged scientists, esoteric women and millionaires – the exotic cast of German cinema of the 1920s. After a few years, visitor numbers began to drop and the planetarium began to look like a short-term fad. The Stuttgart astronomer and planetarium director Robert Henseling wrote in 1931 that 'there was nothing more to be done and the planetarium was threatened with closure.' This complaint that would reappear at various times in the future, such as in the mid-1970s as interest in the skies died as the U.S.–Soviet space race lost popular appeal, and in the 1990s as colour television and cinematic blockbusters seemed more exciting. The Dresden planetarium, which had large numbers of visitors forming long queues in the weeks following its opening, closed in 1933 and became a cinema. The Nazis, who took power in 1933, regarded the whole planetarium venture with distrust, since planetariums supposedly resembled synagogues and were therefore considered part of a Jewish conspiracy. The Nuremberg planetarium, with its clinker facade decorated with sculptures of planets and astrological signs, and which had already been closed due to lack of visitors, was denounced by the Gauleiter Julius Streicher as non-German and pulled down in 1934.

Sadly, most of the other 1920s German planetariums also had a short lifespan. Located in vulnerable city centres, almost all were destroyed or badly damaged in the Allied bombing raids of the Second World War, leaving only Jena and Hamburg unharmed and thus today the oldest still existing planetariums. The great Düsseldorf planetarium was restored in the 1970s and is now used as a popular concert hall, the Tonhalle. The German planetariums of the 1920s had formed a remarkable set of buildings, situated on the ground, on top of towers, in parks, attached to zoos and as parts of grand exhibitions. None of them quite took on the challenge of the minimalist construction of the Bauersfeld dome, but they set a high standard of invention for the planetariums to come. Even if few survived the bombs, they had awakened an enthusiasm for a new building type, which had already been adopted in other countries and which would in the decades following the war spread out across the world.

The arrival of the planetarium was commented on by various critics, including, inevitably, Walter Benjamin. His book *Einbahnstrasse* (One-way Street), published in 1928, is a collection of aphorisms, incidents and observations based on the elements of the modern city. What might one expect to find at the end of the one-way street, but the most recent of building types, the planetarium? The book finishes with the section 'Zum Planetarium' (To the Planetarium). The Berlin Planetarium had opened in 1927, shortly before the publication of the book, and Benjamin added the piece on the planetarium at the last moment. He highlighted the question of a purely optical link to the cosmos, and the replacement of actual experience by simulation:

> Nothing distinguishes the ancient from the modern man so much as the former's absorption in a cosmic experience scarcely known to later periods. Its waning is marked by the flowering of astronomy at the beginning of the modern age. Kepler, Copernicus, and Tycho Brahe were certainly not driven by scientific impulses alone. The exclusive emphasis on an optical connection to the universe, to which astronomy very quickly led, contained a portent of what was to come. The ancients' intercourse with the cosmos had been different: the ecstatic trance [*Rausch*]. For it is on this experience alone that we gain certain knowledge of what is nearest to us and what is remotest from us, and never of one without the other. This means that man can be in ecstatic contact with the cosmos only communally. It is a dangerous error of modern man to regard this experience as unimportant and avoidable, and to consign it to the individual as the poetic rapture [*Schwärmerei*] of starry nights.

If Benjamin was serious in his opposition to observation devices such as telescopes, then few of the astronomical discoveries since the time of the Renaissance would exist, and we would be reduced to the elements of the cosmos visible to our eyes alone. However, his criticism of the artificial nature of the planetarium sky goes deeper. *Rausch* is a term with various meanings, but generally implies an ecstatic trance and was in the 1920s also commonly used to refer to a drug-induced state. *Schwärmerei* implies a lighter state, almost like a mild intoxication. For Benjamin, *Rausch* involves an intensity beyond the simple reverie of *Schwärmerei*, a deep, human feeling of being part of the universe. It cannot therefore be satisfied by an artificial recreation but needs the aura of the original. Bauersfeld, in his proposal for the first planetarium, had described nature as a location from which to see stars, but for Benjamin the term applied to something much greater than simply the natural forms of the planet and was expanded to accommodate new scientific thinking and the elucidation of both the microscopic and cosmic. Benjamin, rather, relies on a myth of an ancient cosmic ecstasy, which he states could only be reached communally. He goes on to suggest a connection between the need for the creation of a virtual optical cosmos – a cosmos in the age of mechanical reproduction – and the social and political disasters of the time, which were partly caused in his view by an overdependence on technology. This technology Benjamin compares to the war machines manufactured and used in the terrible massacres of the First World War during the previous decade. The loss of nature, alongside our ability to celebrate a connection with the greater cosmos, is seen as leading to our predicament today, ever more isolated from the natural world. This viewpoint of Benjamin, who in spite of his reputation as a modern thinker tended to have a fondness for the old and the quaint, may seem regressive, reflecting back to that contradiction of Oskar von Miller as both the bringer of the very electrical lighting which made the night sky invisible and supplier of a replacement artificial version.

One might, however, also wonder about an astronomy and cosmology that is today almost entirely dependent on the achievements of technology, where the current view of the conception of the universe has reached far beyond what can be seen by the human eye. How can we judge remarkable discoveries that rely on the precise performance of wonderful machines, which may also occasionally create their own deceptions and illusions. The Bauersfeld planetarium was still a comparatively simple machine, relying on a fairly basic astronomy of the solar system and fixed stars, which even in the early 1920s was becoming old-fashioned. In 1912, Vesto Slipher in the Lowell Observatory, Arizona – among others – had already observed the redshift in galaxies, implying that the universe is expanding; in 1915, Karl Schwarzschild of Göttingen Observatory had produced the first theory of black holes; and in 1923 the U.S. astronomer Edwin Hubble at the Mount Wilson Observatory in California was observing galaxies beyond the Milky Way. The vast size of the expanding universe was just beginning to be understood, but the planetarium was content, for now, to remain with our solar system and a background of fixed stars.

By rendering the planets and stars as light, rather than as physical objects, by emphasizing beams of the projector rather than the material of the surrounding sphere, Bauersfeld – no doubt unintentionally – contributed to the evolving understanding of the universe as dematerialized, consisting not of objects but of particles and waves. Within the comforting hemisphere of the planetarium there was indeed a feeling of *Rausch*, of looking up and being part of an astonishing communal illusion of being under a celestial vault where distance is uncertain.

The new planetarium was seen by other commentators as educational and entertaining. 'In an age when the moving picture has gained such a hold upon the heart and imagination of our humanity,' wrote Walter Villiger, who acted as lecturer in the first performances at the Jena planetarium, in his book *Das Zeiss-Planetarium* (1926), 'it ought not to surprise anyone that, unable to roam among the denizens of stellar space, we would wish to have what is to be known brought within the reach of our vision by the resources of the star theatre.' Elis Strömgren, director of the Copenhagen Observatory and early visitor to the first dome, wrote, 'It is a school, a theatre, and a cinema in one, a schoolroom under the vault of heaven, a drama with the celestial bodies as actors.' That unlikely dual function of school, where knowledge is acquired but not much fun is to be had, and cinema/theatre, where fantasy and imagination are at home, would remain. Astronomy, which was often seen as a rather dry and uninspiring topic, developed a connection to the entertainment business, producing shows that would rival popular entertainment in the cinema, where in countless sci-fi movies the night sky is a location for thrilling adventures.

Once the lights went down and the spectators' eyes had adjusted to the darkness and the gradual appearance of the stars, the effect of the dome was to give the illusion of being in a much greater space, where distances were hard to judge. In addition, the Jena planetarium created for its spectators the illusion of being on the surface of a spinning planet. Bauersfeld himself described the effect on the spectators as the projected sky began to revolve:

> When the daily movement is switched on, and the heaven of the fixed stars begins to revolve slowly along the polar axis, another illusion appears to the observer, particularly as the movement begins. In the darkness of the room one tends much more to ascribe the movement to the floor on which one stands than to the glowing stars in the sky.

László Moholy-Nagy, photogram, 1925.

As in the cinema, but by different means, the spectators were taken out of their everyday lives and given an unexpected experience, of being in an unfamiliar location.

Weimar Germany's planetarium craze also aroused considerable fascination in artistic circles. Teachers and students from the

Bauhaus, then based in Weimar just 40 kilometres from Jena, were among the first visitors to the Zeiss rooftop planetarium, led by the architect Walter Gropius and the Hungarian artist László Moholy-Nagy. Both would find a natural link between the artistic inclinations of the Bauhaus period, its concern with immateriality and the Bauersfeld invention.

Walter Gropius had been prominent in rethinking theatre performances away from the traditional set-up, with stage and physical sets, towards a more ephemeral atmosphere. Gropius, together with the stage designer Oskar Schlemmer, experimented with projected coloured lights and mechanical ballets. Gropius wrote of the Bauhaus theatre in *Vom modernen Theaterbau*:

> Here is located the cloud apparatus, which can produce clouds, images of stars, and abstract light shows; the projection space, made neutral by the absence of light, becomes through the effects of projected light a space of illusion, the show place for scenic experiences.

Images of stars and abstract light shows: this could be a description of the performance within Bauersfeld's dome. It was not by chance that the dome was named the Sternentheater, and that the astronomer Bengt Strömgren had, admittedly a touch poetically, referred to the stars as actors. Bauersfeld, partly unintentionally, had achieved for astronomy what Gropius was struggling to reach with theatre.

The influence of the rooftop planetarium reached further into the art world. László Moholy-Nagy, who accompanied Gropius on that trip to the Jena rooftop, was fascinated by its lightweight construction techniques and the potential of an architecture created from almost immaterial structures. He saw in Bauersfeld's lattice a potential for a new architecture and placed a photograph of the steel network of the permanent Zeiss planetarium in his book *Von Material zu Architektur* (1929). The geodesic structure of the planetarium dome is seen from below, with the small figures of twelve construction workers suspended in the sky, offering a demonstration of a minimal skeletal architecture without any of the traditional load-bearing elements. The figures, reduced to black shapes against the background of the sky, are held in the net of thin, apparently two-dimensional lines. The effect is very similar to Bauhaus drawings, for instance those by Oskar Schlemmer, where lines of force radiate from human figures as though they are surrounded by a force field produced by the pattern of their movements. Other photographs in Moholy-Nagy's book show the night sky as an inspiration for an architecture composed of light. 'Light', he wrote, beside an image of the stars, 'is a border zone, it creates volume and space.' Moholy-Nagy's interest in forms created from light extended into his numerous photograms, produced throughout the 1920s, before and after the construction of the first planetarium, but becoming markedly more ethereal after his visit to the roof at Jena. In an example from 1925 a white, slightly hazy planet-like form hovers against the darkness of the background. In another, from 1930, a group of black circles hovers above another, larger circle, a cluster of projected forms which appear to be drawn in by the spinning of the larger circle. Many of these photograms resemble the blurry luminescence of forms moving at speed in the skies and have the feel of the shifting circular lights from the planetarium, apparently coupled with the other technique for revealing the invisible – the X-ray.

The great Zeiss projector had an influence on Moholy-Nagy's famous *Light-space Modulator* of 1930, a device constructed of various metal meshes and machines through which lights were projected – a version of the Zeiss projector with an artistic rather than a scientific purpose, producing an early version of a lightshow in which light replaces physical material as the source for art and architecture. A few years later Moholy-Nagy produced a film of the light effects created by his machine, *Lichtspiel Schwarz-Weiss-Grau* (Lightplay Black-White-Grey), with flickering patterns of illuminated shapes and forms revolving and turning within one another, almost a 1930s trial version of the special effects soon to be found in science fiction movies. These photograms and films, often made with domestic objects, combine a feel for the small and everyday with the vastness of space suggested in the planetarium. Through the work of Moholy-Nagy, part of Bauersfeld's legacy spread out from a purely scientific exposition into the light shows and installations of contemporary art. In contrast to Bauersfeld, whose Zeiss system became increasingly concerned with a precise projection system in which the points of light representing stars and planets were rendered as clearly as technically possible, Moholy-Nagy offered a deliberate vagueness, a lack of focus, where the edges of the luminescent forms were fuzzy and unclear and where one is sometimes uncertain what is foreground or background. Art could take liberties where science needed to be exact, but Moholy-Nagy's work was perhaps closer to actual vision, whereby objects are never seen with perfect clearness, especially in the case of the night sky, where the layers of the atmosphere create a drifting luminescence.

The planetarium's influence spread to other artists, and linked up with an already existing interest in depictions of the stars and planets. In 1931 Man Ray produced a photogravure of a planet with an electric switch, as though the planet could now be turned on and off. The American sculptor Alexander Calder, when living in Paris in the late 1920s and already associated with mechanical theatres, created a series of artworks based on the theme of spheres and circles, resembling minimalist orreries. Calder visited Berlin in 1929, and probably saw the new planetarium. Calder loved the irregular motion produced by the unexpected functioning of the planetarium projector as much as the steady paths of the planets, his version of the universe was full of unexpected moments. He would write in his essay 'What Abstract Art Means to Me' (1951) that

> the underlying sense of form in my work has been the system of the universe . . . the idea of detached bodies floating in space, of different sizes and densities, perhaps of different colours and temperatures, and surrounded and interlarded with wisps of gaseous condition, and some at rest, while others move in peculiar manners, seems to me the ideal source of form . . . A very exciting moment for me was at the planetarium, when the machine was run fast for the purpose of explaining its operation: a planet moved along a straight line, then suddenly made a complete loop of 360 degrees off to one side, and then went off in a straight line in its original direction.

Calder's previous interest in the balancing acts of acrobats in the circus ring was transferred to the gravitational systems operating between celestial bodies, which hold one another in balance. *Two Spheres within a Sphere* and *Black Spot on Gimbals*, both from 1931, resemble solid planets caught within a system of influences, suggested by the wires. *A Universe* (1934) was a sculpture of twisted wires along which a red sphere and a white sphere moved slowly, propelled by an electric motor. Albert Einstein is reported to have watched the movements of these spheres for a long period, meditating perhaps on their relative movements. These early Calder pieces resemble more Oskar von Miller's other planetarium, with its models of planets, than Bauersfeld's illuminated show, but the inspiration to make these devices was part of a wider fascination with astronomy paralleling the development of the planetarium.

And as for Bauersfeld himself, the man who created the private sky: what kind of person was he? Bauersfeld lived in the shadow of his great invention, and today is hardly well known. He was a technocrat, but also a visionary, who lived through difficult times. He was born in 1879 in Berlin, the son of a shoemaker, and lived in considerable poverty. As a boy his dream had been to become an astronomer, but he had trained as a machinery technician. Employed as a workshop technician at Zeiss, he worked at first on lens making, then on cinema projectors and optical devices, and was known for his finesse with delicate machinery. The combination of these skills enabled him to design the planetarium projector, which required a knowledge of lenses and lighting, the ability to evolve refined moving parts, and an understanding of the transference of two-dimensional images onto the three-dimensional dome. In a photograph from 1925 he can be seen leaning against a wall, dressed rather elegantly in a long jacket, with his wife Elizabeth and his large family of five girls and three boys. He has clearly moved up from being a lab technician. In photographs from the later 1920s we see a respectable man, of unremarkable appearance, dressed correctly in a dark three-piece suit. He appears in various group photographs from the Zeiss archives, among his colleagues, but never stands out. In the 1930s, certain shadows begin to appear. The political changes brought about by the rise of the Nazis meant that the directors of important industrial works could no longer remain neutral. Bauersfeld can be seen, as one of three directors of Zeiss, standing beside uniformed Nazi members, surrounded by swastika flags – even with his arm raised in the Hitler salute. In 1937 he became a member of the Nazi Party, possibly from conviction, possibly as a move made necessary by his senior position in the firm.

The Nazis, despite a considerable desire to control the skies, had little interest in planetariums, and none were built during their period of control of Germany. The first period of mass enthusiasm for displays within the planetariums faded as the political conditions in Germany darkened. However, the influence of the

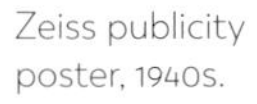

Zeiss publicity poster, 1940s.

Bauersfeld projection system can be seen in Albert Speer's 'cathedral of light', used at the Nuremberg Rally of 1934, when a wall of 152 searchlights were turned upwards into the sky. As in the planetarium, light was projected vertically, but now into the real rather than artificial sky, for political rather than scientific purposes, for a mass audience united by political fervour rather than one seeking entertainment. Later variations converged the searchlights to one point, over the heads of the spectators. 'The feeling', wrote Speer in his autobiography, 'was of a vast room, with the beams serving as mighty pillars of infinitely light outer walls.'

The production of the Zeiss works was turned increasingly to the optical aspects of weaponry – bomb and gun sighting technology, tracking devices and so on. With the shortage of manpower, large numbers of prisoners and forced labour were introduced to keep the factories in operation. Bauersfeld, as a director of Zeiss, had the responsibility of supervising this slave labour force. The night sky, which Bauersfeld had created as an artificial projection, formed in the early 1940s the actual location – partially illuminated by Zeiss searchlights – for the Allied bombers, bringing destruction to the German cities, including most of the planetariums with their fragile domes. At the end of the war American troops arrived in Jena and took over the Zeiss works, knowing that the city would shortly become part of the Soviet zone. Bauersfeld, along with most of Zeiss's directors and upper-level scientists, was quickly seized and, in spite of his opposition, dispatched to Heidenheim an der Brenz, in southwest Germany and in the American zone. In the nearby town of Oberkochen, the Americans established the reconstruction of Zeiss as a West German firm, producing high-quality optical equipment. Photographs from this time show Bauersfeld as an elderly figure, still wearing the same dark three-piece suit, seated at official meals in Swabian restaurants, surrounded by other Zeiss executives, and at his seventieth birthday cutting a multi-tiered cake on which is written a large W. He died in 1959, an honoured but already rather forgotten figure. The great dumbbell projectors nevertheless continued to throw out their beams of light, and to allow the projected planets to pass across the night sky, the flickering comets, created by ingenious lenses and usually considered linked to some unspecified slightly threatening event, to course at speed across the surface of the ingenious dome he had invented.

THREE

RED STAR, WHITE STAR

The star theatre had become a success. Affluent businessmen arrived from the USA, revolutionaries from the USSR, professors from Argentina, astronomers from Scandinavia, writers and artists from France, politicians from Japan. Following the popular acclaim in Germany of the Wonder of Jena, the reputation of the planetarium spread rapidly, as these visitors travelled to Germany to experience the new artificial sky for themselves.

Several cities in Europe, the U.S. and Asia constructed their own planetariums, using, with a few notable exceptions, the Zeiss Mark II dumbbell projector devised by Walther Bauersfeld and Walter Villiger to show both the northern and southern hemispheres, which was continually improved during the 1920s to project an increasing number of stars and ever more special effects. In 1927 Vienna put up a temporary timber building near the Prater funfair. The following year a planetarium dome was constructed in Rome, inside the Octagonal Hall of the Baths of Diocletian, a classical ruin from the Roman Empire that Italian culture of the period considered the natural location for investigation of the new imperium of the heavens. In 1929 Moscow built a planetarium, as did Stockholm, Milan and Chicago – the first in the USA. A permanent building, reusing the projector from the temporary structure, was constructed in Vienna in 1930. Philadelphia followed in 1933; The

ЛЕВ
ГОНЧИЕ
ПСЫ
ВОЛОСЫ
ВЕРОНИКИ
ДЕВА

Hague in 1934; Brussels, Los Angeles and New York in 1935; the bizarre San Jose Rosicrucian Park planetarium was opened in 1936. Paris installed a planetarium within the Palais de la Découverte in 1937. The first Asian planetariums opened in Osaka in 1937 and Tokyo in 1938. Pittsburgh completed the sequence in 1939. After this, the flow of planetariums paused for over a decade, interrupted by the struggles of the Second World War as the skies were used for rather less peaceful purposes.

Some of these planetariums, such as those in Paris, Rome and Philadelphia, were simply hemispherical screens within existing buildings, and therefore have no independent architectural exterior. Others were domes on rooftops in the Jena tradition, such as those in Osaka and Tokyo, and thus also have little individual architectural identity. But many were independent buildings and therefore required an architectural form. The nature of this form varies and produces a fascinating and very diverse set of buildings, reflecting the concerns of those who commissioned the buildings and the social and political background of their location. The most challenging of these, and in many ways the most interesting planetarium of all, was opened in Moscow in 1929, under the socialist government. The idealism of the Moscow planetarium can be contrasted with a series of U.S. planetariums built in the 1930s, which originated from a wide range of intentions.

IN THE SKIES OVER MOSCOW, in the decades before the collapse of the socialist state, stood three symbols of the space programme: the rocket, the cosmonaut and the red star. The rocket is still atop the Monument to the Conquerors of Space (1964), a 110-metre-high titanium sculpture beside Prospekt Mira with the Alley of Cosmonauts leading to its base. The cosmonaut Yuri Gagarin stands on a 30-metre-high column on Leninsky Prospekt, his arms pulled back in the style of a classic Marvel superhero, as though about

Moscow Planetarium, schoolchildren with celestial sphere, 1950s.

to leap upwards towards the stratosphere. These two monuments look at once back to the period of Soviet space exploration and forwards to the time of planetary probes and space stations. The red star, symbol both of astronomy and of communism, preceded the epic period of space flights and once crowned the dome of the Moscow planetarium located on Sadovaya-Kudrinskaya street.

This planetarium stands at the intersection of influences created by politics, engineering, style, theatre, astronomy, space exploration and religion, each of which affects the others. It was one of the last buildings to be put up in the style known as Constructivism, and thus looks both back to the original fervour of new Soviet society, and forwards to the difficult decades of state socialism.

Constructing a planetarium in Moscow was never going to be merely about an educational display of the movements of the planets and the stars, however inventively demonstrated by the Bauersfeld projector. Russia had a longstanding and diverse attitude to the cosmos, mixing science, esotericism, Gnosticism, a belief in a world beyond the purely physical, and a feeling that the nation was specially destined to explore the planets and beyond. The nineteenth-century writer Nikolai Fyodorov, founder of what is known as Cosmism, had proposed that the atoms of the deceased were scattered throughout the universe, and that steps should be taken to resurrect the dead, who would then live on other planets, since there would not be enough room for them to inhabit the Earth. This interplanetary resurrection theme would later filter into Soviet science fiction, for instance in Andrei Tarkovsky's film *Solaris* (1972), where the dead reappear on a spaceship above an aquatic planet. However, Fyodorov also inspired early Russian ideas on planetary travel. Konstantin Tsiolkovsky, the first Russian rocket designer – who in Tsarist times had already worked out the basic formulas for the thrust required by engines intending to escape the Earth's atmosphere – had been a follower of Fyodorov.

Tsiolkovsky constructed the first models of rockets and dirigibles, indulged in complex and semi-mystical theories as to the make-up of the universe, and wrote science fiction novels about Russian space explorers encountering aliens on other planets. He was deeply eccentric – photos show him with long, flowing locks, surrounded by his rockets and airships as he holds up his ear trumpet as though to detect distant sounds.

Early Russian science fiction writers described the coming age of interplanetary flight. Alexander Bogdanov in *Red Star* (1908) described Mars, the definitive red planet, as inhabited by a benign race keen to associate with the Russians. Aleksei Tolstoy wrote the novella *Aelita* (1923) at a time when the colour red had acquired political significance. It tells of two Soviet cosmonauts who visit

Moscow Planetarium, exterior, 1929.

Mars, discovering that its population is a slave race dominated by corrupt rulers and priests and assisting in a 1917-style revolution, made slightly more complicated by the lead cosmonaut falling for the Martian queen. *Aelita* was made into a spectacular and very popular film by Yakov Protazanov in 1924, with remarkable Constructivist-inspired sets.

Societies with mass membership were set up in Moscow in the 1920s to promote a Soviet flight programme. The popular Soviet enthusiasm for all things planetary inevitably required the construction of a planetarium. There was considerable interchange between Weimar Germany and the socialist USSR; both saw themselves – however misguided it would later turn out to be – as new nations with an interest in creating a new world based on a broad popular culture, and in the 1920s both had highly progressive artistic, theatrical and cinematic movements. Rumours of Bauersfeld's invention quickly spread to Moscow. David Riazanov, a former comrade in exile of Leon Trotsky and director of the Marx-Engels Institute, an organization devoted to Soviet philosophy and history, proposed the construction of the planetarium in 1926. The building was originally intended to be part of a large science complex – including a zoo, a museum and a library – that would exemplify the rising power of pure science. Evolutionary time would be presented in the zoo and cosmic time in the planetarium, both set against the traditional religious time of the discredited Russian Orthodox Church. Riazanov travelled to Germany to visit various planetariums and to commission Bauersfeld and the engineer Franz Dischinger to construct a planetarium for Moscow. Bauersfeld and Dischinger possessed skills lacking in the USSR: the technology to construct the projector and the engineering expertise to put up the ferro-cement dome. Lightweight construction and the projection of light would thus be combined to produce a building of the new age, with the minimum use of material. This Moscow planetarium was to be on an ambitious scale, with an internal diameter of 27 metres and seating for 1,440 spectators. The Zeiss dumbbell projector provided by Bauersfeld was the latest model, now able to project 8,956 stars and to switch the point of projection to various latitudes, a useful facility considering the vast size of the USSR.

The architectural design of the planetarium is credited to two young Russian architects, Mikhail Barshch and Mikhail Sinyavsky, and is usually referred to as Constructivist – though Constructivism included any number of different approaches, and often implies little more than modernism. While both architects had emerged from the Vkhutemas school of art and architecture, established in 1920 by Lenin, they were also conversant in the eclectic mix of styles left over from the pre-revolutionary period. The planetarium was the first, and only, building by this pair of architects. A photograph from the time, which looks to have been taken in late evening, shows the architectural team climbing up the maintenance ladder attached to the outside of the dome with Barshch in front, as though leading his colleagues into the skies. Asked later how they had come to be awarded this prestigious commission, Barshch commented that nobody in Moscow knew what a planetarium was – it was presumed to be some kind of toy for children, and therefore more experienced architects had not been interested in the project.

The Moscow Planetarium was thus a mix of German optical and engineering technology, which determined the basic layout of the building, and Soviet modernist architecture, which supplied a specific external form. It was built at a time when the experimental culture of the revolutionary period was coming to an end and in the early years of Stalinism a mix of modernist and traditional forms had become the official style. In fact, the various designs for the planetarium flickered between modern and traditional. An intermediate design by Barshch and Sinyavsky proposed a large neoclassical porch with Greek columns, looking back to pre-revolutionary architecture as well as forwards to the Stalinist period, and reminiscent of the neoclassical planetarium in the Jena park. This concept was abandoned for the clean, simple form actually constructed, but hovers around it as an unconstructed phantom project. In fact, the Moscow building was the only permanent planetarium of the period to really take on the notion of being an uncompromising modernist building.

Much of the design of the Moscow Planetarium was determined by the technical requirements of the Zeiss projection system. The building has three storeys: a basement, an entrance and foyer at ground level with a set of radial portal frames to support the floor above, and a main projection hall on the first floor. Additional service spaces needed to be added to the circular form, so four elements project from the central circle – the entrance, with its vertically curved wall; a volume for storing the projector to the side when it is not in use; another volume with an elegant, glass-clad spiral staircase; and offices for the staff. These four protruding volumes create a dynamic plan, at once circular and also pushing out from the centre to the periphery.

The exterior dome, constructed in ferro-cement, was given a parabolic form, which was highly unusual – in fact, it was the only parabolic dome ever built over a planetarium. This outer dome is only 8 centimetres thick at the top and 12 centimetres at the bottom, achieving a ratio of shell thickness to internal volume of 1:280 – less than that of an eggshell to an egg. It was constructed from a framework of steel rods laid according to a timber formwork and onto which was sprayed concrete. Due to a shortage of materials in Moscow, a cement substitute was created from ground clamshells, which have roughly the same chemical composition as cement – and which for some reason were available in landlocked Moscow. The insulation was a layer of moss. The materials of the sea and of the land were used to create a shell for the artificial sky, which was then covered externally with aluminium sheeting imported from Germany.

There were precedents for parabolic domes in Germany: the domes of the unbuilt Meyer planetarium, and of the Glass Pavilion of Bruno Taut – constructed in Cologne for the 1914 Deutscher Werkbund Exhibition and which the German writer Paul Scheerbart described as allowing 'contemplation of the light of the moon and stars'. But there were other Russian and Soviet influences. Mikhail Barshch had also been on a tour of Soviet Asia and was interested in the diverse forms of mosque domes, some of which are parabolic. Parabolic domes were becoming fashionable in Moscow, appearing in Ivan Leonidov's visionary project for a Club of the New Social Type (1928) as well as Moisei Ginzburg's unbuilt proposal for the Palace of the Soviets (1931). From the exterior, the planetarium resembles a cylindrical egg cup cradling a great egg, which is particularly significant to the Russian celebration of Easter, the time of rebirth and resurrection – the linking of the daily rising of the Sun and the rebirth of the soul is a reminder of a theme that goes back to those images of the Egyptian goddess Nut. The red star at the summit of the dome inevitably evokes the stars on traditional Russian onion domes. It is not necessary to say that one of these influences was decisive; a building can be indebted to diverse, even contradictory sources.

One of the themes of certain early Soviet artists, such as Kazimir Malevich, had been the elimination of the object and its replacement by abstract shapes and lines of movement. The Moscow Planetarium can be seen as constructed of various movements. In section, the parabola of the exterior dome rises above the semicircle of the interior projection screen, the parabola outlining the elliptical path of planetary motion and the semicircle the more ancient notion of the perfect sphere of the heavens. The curves within the radial portal frames on the ground floor and the partial parabola of the entrance canopy echo these lines, so that the entire section consists of curved lines of potential movement. Meanwhile, the circular plan of the planetarium is based on both circumferential and centrifugal movement, as the circular lines of the great circular hall are balanced by the four forms flung outward, including the elegant stair spiralling in its glass cylinder. The architectural dynamics determine the movement of the spectators, who enter axially through the curved entrance, circulate among the portal frames that radiate out from the centre of the foyer, and move up along the steps running around the circumference of the building to take their places in the projection hall. To these can be added the lines of planetary movements created by the Zeiss projector. The illusion that the dome of the building has vanished and that there is only a night sky above results in the planetarium's final dematerialization.

In 1927 Aleksei Gan, editor of the architectural magazine *Sovremennaya Arkhitektura* (Modern Architecture) and Constructivist theorist, published drawings of the first version of the planetarium and proclaimed it to be a successor to both traditional Russian theatre and the Orthodox Church:

> The theatre has been up until now nothing but a building dedicated to the service of cult. How this service has been performed, to which cult it is dedicated, plays no role . . . Our theatre must be different; it should draw the spectator to a love of science. The planetarium – a theatre of optical science – is also one of the forms of our theatre. In it, people do not act, but manage a complex technical apparatus. In this theatre, all is mechanized . . . So the theatre at the service of cult passes to the service of science. In this theatre, Man, equipped with machinery, extends his sense of perception, sees the most complicated mechanics of the movement of celestial bodies. This

> will help him forge a scientific understanding of the world and free himself from both the fetishism and prejudice of savage priests and the pseudoscience of European civilization. For this theatre, we need to build a new building.

For Gan, traditional, pre-revolutionary theatre was no more than a kind of church service, whereas the planetarium would produce the most refined version of the theatre performances proposed by Vsevolod Meyerhold and Lyubov Popova, in which narrative was abandoned, actors moved in precise patterns, and sets became large machines to accommodate these movements. But by 1927 experimental Soviet theatre performances were already part of the past, as the original impetus for a new theatre had faded. For those like Gan, still hoping for a revival of the earlier enthusiasm, the planetarium offered a new opportunity. Its form resembles a temple, within which the audience sits reverently as the mysteries of the heavens are revealed on the interior of the dome – the traditional location in Orthodox churches for a large painting of God staring back down.

Other members of the Soviet cultural vanguard were fascinated by the planetarium. Alexander Rodchenko visited the building several times and produced a series of photographs in his characteristic style, with the camera held at an angle, giving the images an unsteady feel, as though the surface of planet Earth has somehow shifted off its axis. He recorded the dome in the snow, the Zeiss projector, the extruded service boxes, and a suitably cloth-capped proletarian descending the spiral staircase; a similar figure features in his shots of the Shukhov radio tower. Some of these photographs were featured in Gan's article in *Sovremennaya Arkhitektura*. Rodchenko wrote in his autobiography *Black and White* (1939):

A planetarium appeared in Moscow.
This was an enormous fantastical apparatus.
It was the realisation of his fantasy.
Made of black metal and glass.
With forms that resembled no living creature.
It was called The Martian . . .
It made him search and search for a fantastical reality.
Or for the fantasy in reality.
And to show a world that people had not yet learned to see, from new perspectives, from points of view, and in new forms.

The planetarium is here linked to Mars, the red planet, and to fantasy aliens – H. G. Wells's *War of the Worlds* was very popular in Russia – and new forms of life. Rodchenko suggests that the task of the photographer and of the planetarium were linked. The planetarium offered a new vision of the cosmos that no one had seen before, and this inspires him to look in a new way through the camera lens, to take photos that reveal previously unobserved aspects of everyday life.

This interest in the planetarium as a revolutionary building was shared by the poet Vladimir Mayakovsky, a close colleague and friend of Rodchenko, who may have visited the planetarium at the same time as the photographer. Mayakovsky was another mesmerizing early Soviet figure, at times too individual to handle and at others a kind of bootboy for the revolutionary state. In one of his final poems, written shortly before his suicide in 1930, Mayakovsky declared, in his characteristically individualistic grammar:

Proletarian woman
proletarian man
come to the planetarium.

Moscow Planetarium, stairs.

Come in
hear the lively buzz.
In the lecture hall.
Spectators sit awaiting the sky to be shown
The head-sky-manager comes
the expert in sky matters
He comes
pushes and twirls the million celestial bodies.

The proletarian masses become charged with some invisible celestial force. The 'head-sky-manager' – the lecturer who also controlled the projector – becomes in Mayakovsky's words part scientist, part priest, part shaman and part theatre director. As is often the case with Mayakovsky, his support for the new socialist world has an undertone of mysticism: the planetarium is there to educate the masses, but it is also mysterious, magical. The projector could speed up time, summon up vast cosmic spaces, and move at will its viewing point to a distant location. It was an appropriate mechanism to revive a declining revolutionary impulse.

The large crowds that came to the Moscow Planetarium to see the scientific display of the paths of the planets and stars understood that the shows were also intended as an inspiration for the future expansion into space of the Soviet people. Tsiolkovsky's model rockets, accompanied by images of Soviet colonization of other planets, were shown prominently in the lobby of the planetarium. Two rockets were placed by the entrance, imitating columns, as though architecture and rocketry could be combined. In the early 1940s the Moscow Fantastic Theatre put on shows about Copernicus, Giordano Bruno and Galileo. During the early 1960s, Soviet spacecraft, such as the spherical *Vostok 3KA-2* launched with the space dog Zvezdochka shortly before the Yuri Gagarin mission, were displayed in the lobby on the ground floor of the planetarium. In 1957 a large globe was set up on the ground floor of the planetarium, on which was plotted Sputnik's changing position in orbit above the Earth. Visitors would be inspired by the gradual expansion of the Soviet programme beyond the Earth's atmosphere. The planetarium was used in the early 1960s to show the workings of the solar system to prospective cosmonauts, some of whom, such as Gagarin, returned to lecture on their experiences to large audiences. The planetarium was thus linked with the early achievements of the Soviet space programme – the visible sign within Moscow of the success of Soviet technology in comparison to its Western rivals.

The later story of the Moscow Planetarium mirrors that of the Soviet Union. It had been brought to life at a moment when Soviet society was rapidly changing, when the great hopes of the revolution were fading before a political system moving towards the Stalinist dictatorship of the 1930s. In the 1930s the Stalinist government added the red star – it did not feature in the original design – so as to link the planetarium to the regime. At some point, the exterior walls were painted bright blue, as though to deny its links with the white modernism of the 1920s, and the building's curving entrance porch was removed. A photo from the early 1940s shows the building already in a decayed state, surrounded by anti-aircraft batteries dug into the rubble around the building, ready to fire up into, rather than investigate, the heavens. After the war the Soviet space programme provided a new sense of purpose for the planetarium; it was renovated, received a new state-of-the-art Zeiss projector in 1970, and once again became a popular feature of Moscow society.

By 1987 the planetarium was listed as a monument, but with the collapse of the Soviet Union, the building soon fell prey to the uncertainties of the times. With the change in the political system, the red star was removed and replaced with the Russian flag, and the institution privatized. The valuable site was situated in an affluent residential area, and rival property developers struggled for ownership through legal and illegal means. Armed thugs raided the building and threatened the staff, and some of the objects in the foyer were stolen; some remain missing, others were hidden by the loyal staff. Finally, the site was sold to a development company. The building remained closed for several years, during a period of political struggles, corrupt judges and bankruptcy. The revival in the fortunes of the Russian state, however, coupled with the public enthusiasm for the new age of Russian space exploration shown by the Mir and International space stations, eventually led to redevelopment of the site. Just as in the USA and Europe, where early planetariums were seen as outmoded, the Russians felt in the late twentieth century that the Moscow Planetarium was no longer worthy of their ambitions in space. The new plan called for the construction of a large-scale, four-storey astronomy museum designed to accommodate not just a planetarium but a cinema, lecture rooms, museum exhibits and educational displays. Most of the old planetarium building was, controversially, destroyed. Only the cupola, the portal frames and some other minor elements were preserved; the rest of the original structure was rebuilt, and when the reconstructed planetarium was finally placed on top of the new museum building, it occupied a position six metres higher in the Moscow sky than it had previously. What stands there today looks something like the old building, but without much of its original atmosphere.

In 2011 the new planetarium, equipped with the latest Zeiss digital projector, opened to the public. Today the building is part of a complex that includes a small park containing a selection of astronomy-related objects, two small observatories, models of Stonehenge and Jaipur, and glass pyramids and spheres: a broad version of astronomy mostly shorn of its specifically Russian themes. The hopes of Russian space exploration, kept alive by the existence of the International Space Station, have recently shown signs of revival, with declarations by Vladimir Putin that the colonization of space is once again the destiny of the Russian nation.

After the success of the planetarium, Mikhail Barshch would progress to a long and successful career in the USSR, with projects ranging from avant-garde housing schemes to Stalinist neoclassical projects, adapting to the severe constraints of the times but always with a certain individual style. In his final years Barshch was one of the architects for the Monument to the Conquerors of Space, with its soaring rocket, balancing his earliest building at the beginning of Soviet rocket science with the new museum dedicated to the achievements of the space programme. Both buildings are concerned with looking up to the heavens, which await exploration by cosmonauts.

At the beginning of Aleksei Tolstoy's *Aelita*, the egg-like spaceship raises itself a few metres above the city before beginning its surprisingly rapid journey to Mars. The egg-shaped dome of the Moscow Planetarium, in a state somewhere between reconstruction and resurrection, both ghost and portent, has also now been elevated, as though it too were about to begin a longer flight. Once again, it takes its place alongside Yuri Gagarin on his column and the rocket on its vast plume, all three in a state of barely restrained stasis, awaiting their next skyward move.

THE CONTRAST AS ONE SHIFTS from the USSR to examine planetariums in the USA is revealing. While the Soviets had Fyodorov and Tsiolkovsky, and later the cosmonauts Yuri Gagarin and Valentina Tereshkova, the U.S. had developed its own culture of science fiction and space travel. From the late 1920s, sci-fi magazines like *Amazing Stories*, *Astounding Stories* and *Thrilling Wonder Stories* encouraged the idea that space travel was not far off and that Americans would soon be roaming on other planets. The titles speak for themselves: 'Slave Raiders from Mercury', 'Giant Computer Rides the World', 'Laboratory of the Mighty Mites', 'Crusade across the Void', 'The Yellow Men of Mars', 'Beyond the Rings of Saturn', 'Tongues of the Moon', 'Swamp Girl of Venus' – and in the same astonishing issue 'The Vibrator of Death'.

This fictional version of space, which would have a considerable impact in raising interest in a version of astronomy always hovering between actual and imagined, was paralleled by more straightforward scientific developments as astronomers continued to explore the cosmos on an ever greater scale. In the early 1920s Edwin Hubble, working at the Mount Wilson Observatory, confirmed that our galaxy is only one of many and also that the universe is indeed expanding; in 1931 the Belgian priest Georges Lemaître, whom many think preceded Hubble with the notion of the expanding universe, produced the first theories of the Big Bang, which he named the Cosmic Egg; in 1930 Walter Baade and Fritz Zwicky theorized about neutron stars and Clyde Tombaugh discovered Pluto, thus confusing both planetarium projector manufacturers and astrologers as to just how many planets should be taken into account. In 1932 Karl Jansky of the Lowell Observatory first detected radio waves from within the Milky Way, leading the amateur radio operator Grote Reber to build the first parabolic sheet radio telescope in his back garden in Wheaton, Illinois; in 1938 the Jewish German physicist Hans Bethe, working in exile at Cornell University, published the first description of nuclear fusion in stars.

It took a while for the significance of these remarkable discoveries to filter through into popular science and to become relevant to the planetarium, the relatively simple view of the night sky from the surface of the Earth being difficult to adjust to the idea of there being vast numbers of galaxies expanding into an indefinable space. Astronomy in the U.S. was not linked to revolutionary fervour as it was in the USSR, and therefore the architecture of the planetarium did not require any equivalent form of Constructivism – any sign that the planetarium was part of a progressive contemporary culture. The various planetariums constructed in the USA in the 1930s reflected the culture of American society of the time – capitalist, religious, expansionist and often self-contradictory. Popular science was a way of understanding the world, and was linked to mass entertainment, to adventure, exploration and individual encounters with the unknown. American planetariums were sometimes eccentric, combining pursuit of the latest scientific discoveries with a fascination for ancient religions, showmanship and a concern for moral improvement, and individualistic and mass popular culture. These apparently diverse interests often sit comfortably enough together, and led to a diverse range of architectural proposals.

The first American architect to see the potential in the Bauersfeld planetarium was Frank Lloyd Wright, who included a planetarium in his proposal for a resort on the peak of Sugarloaf Mountain, Maryland, under the patronage of the Chicago real estate tycoon Gordon Strong. This project, known as the Gordon

Frank Lloyd Wright, project for Sugarloaf Mountain planetarium, 1925.

Strong Automobile Objective, dates from 1925, barely two years after the opening of the planetarium on the Jena roof and before any permanent planetarium had been constructed in Germany or anywhere else. Wright probably heard about the Jena planetarium from the German architect Erich Mendelsohn, who visited Wright in the USA in 1925. Mendelsohn had just completed the Einstein Tower in Potsdam, an astrophysical observatory that housed complex equipment manufactured by Zeiss for measuring the spectrum of solar rays in order to test out certain aspects of Einstein's theory of relativity, and which combined advanced scientific research with expressionist architecture. Gordon Strong owned Sugarloaf Mountain, and wanted a building to crown the summit as a destination for the increasing number of automobile owners. Strong was uncertain as to quite what would provide the attraction and considered a nightclub or picnic centre, while Wright proposed a planetarium as being more appropriate to the location and the spirit of the times – cars would be combined with stars. Wright's final version of the project consisted of an external double-spiral ramp with separate roadways one over the other for the cars to ascend and descend. Within the conical interior of this spiral would be the planetarium. Cars would spiral around the inside in the open air, as planets followed their projected paths across the night sky within. In Wright's sectional drawing this planetarium resembles Bauersfeld's hemispherical dome and central projector. The Wright planetarium dome would have been similar in construction to the thin shell concrete of Jena, but with a diameter of 50 metres would have proved much too large for any projector of the period. In any case, Strong had little interest in astronomy and found Wright's high-minded proposal impractical, and the project was soon abandoned.

Wright never attempted another planetarium but the spiral theme returned: turn the Sugarloaf Mountain project upside down

Queues outside the Adler Planetarium, Chicago, 1930.

and it becomes his Guggenheim Museum in New York, with the pedestrian ramp winding downwards and inwards, the dome of the planetarium replaced by a shallower form letting in actual daylight rather than the projection of an artificial night sky.

Four years later another Chicago tycoon picked up on the notion of building his own planetarium. Max Adler was a German Jewish businessman who had originally been a concert violinist but had given up his musical career in favour of the acquisition of great wealth, becoming vice-president of the department store chain Sears Roebuck & Co. In the late 1920s he abandoned his considerable business interests for philanthropy. He travelled to Germany with his cousin, the Chicago architect Ernest Grunsfeld Jr, visited the Munich Deutsches Museum and the Bauersfeld planetarium, and was filled with enthusiasm for the new invention.

Adler hired Grunsfeld to construct the first planetarium in the U.S. on an, at the time, isolated site along Lake Drive, beside Lake Michigan. He ordered from Zeiss the latest version of the Mark II projector. Grunsfeld was interested in the Mayan architecture of ancient Mexico and stepped pyramids. His planetarium, clad on the outside with rainbow granite, has a stepped form, with a series of concentric walls built to a fascinating twelve-sided plan, and crowned by a copper-clad dome. The building is another planetarium temple, following the tradition established in Germany but now with a form vaguely reminiscent of ancient Mexico. Each

Adler Planetarium, 'The Drama of the Heavens', 1939.

Adler Planetarium, Chicago, show, 1930.

of the twelve walls feature sculptures of a sign of the zodiac, linking the building both to earlier astronomy and to astrology. The concentric rings give the impression to the visitor of entering, like an initiate to a cult, through a series of stages, until arriving at the inner sanctum of what was known as the planetarium chamber.

Adler spoke of his notions of the cosmos at the opening of the planetarium:

> The popular conception of the universe is too meagre; the planets and stars are too far removed from general knowledge. In our reflections we dwell too little on the concept that the world and all human endeavour are governed by established order, and too infrequently on the truth that under the heavens everything is interrelated, even as each of us to the other.

The planetarium was to be more than just a demonstration of the solar system: it also had a philosophical and spiritual purpose – the holy theatre could still thrive in the age of scientific discovery and combine with the rough, or popular, theatre. Philip Fox, later director of the Adler Planetarium, wrote that the planetarium had a religious purpose and that the visitor might 'wander through the depths of space and measureless time and touch the divinity that broods there'. Who was this brooding divinity? A god who had been edged out to the periphery of his own heavenly show, but who remained, just out of reach of the projector beams, as a distant prime mover. And if the Adler Planetarium, clearly scientific in intent, is based on the form of a pagan temple, one might wonder whether there is just a hint here of the performance of a repeated ritual of the holy theatre, not just to show the heavens, but also to influence them, to ensure the planets continue to rotate.

The Adler Planetarium attracted considerable popular attention, drawing long queues of visitors to see the shows. Unlike German planetariums, which tended to repeat the same programme, the Adler operated more like a cinema, with a sequence of changing programmes illustrating different aspects of the universe, so visitors would return to see the new show. Adler also acquired a large collection of historical astronomical instruments from a German collector, such as astrolabes, orreries, telescopes and sundials, which were placed in a rather haphazard fashion in the two exhibition rooms within the building. Visitors passed through these rooms on the way to the planetarium chamber, the history of astronomy leading up to its current achievements.

In 1997, long after the patron's death, the Adler Planetarium also acquired the Atwood Sphere, the revolving sphere that demonstrated the starry heavens, built in 1913. Shortly afterwards, in a move typical of the period as new achievements in space made the technology and architectural design of old planetariums seem old-fashioned, the Adler expanded, with the construction of a large semi-circular glass building enclosing the original Mayan temple. This space houses a state-of-the-art digital projection area, linked by a moving walkway to the original planetarium chamber. There are thus now three planetariums from different periods: the Atwood Globe, the planetarium chamber with a now updated Zeiss projector, and the space with the digital projection system showing contemporary astronomy. The three different forms of architecture – steel globe, granite dodecahedron temple and high-tech semi-circular building – reflect changes over time to both astronomy and architecture. Still, something remains of the quiet, almost meditative idea of the mysterious temple under the stars beside the lake.

Other affluent philanthropists followed Adler's example. In 1933 the household soap tycoon Samuel Simeon Fels, also of German origin and another devotee of classical music, constructed the Fels Planetarium inside the Franklin Institute in Philadelphia. Since the planetarium is set within another building it has no architectural exterior.

Other tycoons were more determined to see their philanthropic activities enshrined in a visible architectural monument. The spirit of King Khosrow and his dome of the skies constructed to emphasize his royal power survived into capitalist USA. Charles Hayden, a New York banker who had made a fortune from investing in copper mines, producing metal for the booming electric wiring industry, was the patron of the Hayden Planetarium, which opened in New York in 1935. Hayden was a true philanthropist in the U.S. tradition of J. P. Morgan and Andrew Mellon – robber barons turned good – donating large sums during his lifetime to hospitals, organizations for moral improvement and cultural institutions, but also ensuring that his name was linked to the size of the donation. *Fortune* magazine commented that 'Mr Hayden

Hayden Planetarium, New York, postcard.

believed that everyone should have the experience of feeling the immensity of the sky and one's own littleness,' a feeling of size that may vary depending on whether one's place in American society is as pauper or tycoon.

The Hayden Planetarium of 1935 was attached to the American Museum of Natural History and designed by the New York architects Trowbridge & Livingstone, a firm producing stolid Beaux Arts buildings. They produced a rectangular brick volume with a colonnade at the entrance and a bronze dome – a serious-minded civic building. The location beside the Museum of Natural History offered an interesting take on the idea of the planetarium creating an artificial version of nature. It would sit beside not only the sequence of rather melancholic halls filled with stuffed animals, cephalopod skeletons, fossils and meteorites – the lifeless detritus of the natural world – but dioramas such as the Akeley Hall of African Mammals, where animals stuffed by the taxidermist explorer Carl Akeley were placed in scenes representing their natural habitat – nature theatres. These dioramas created the illusion that the creatures were still alive, in their natural habitat, even if frozen into a motionless state. Shows of illuminated artificial stars and planets would now balance the stuffed gorillas in the jungle and the elephants on the plains.

The interior dome of the Theatre of the Sky, onto which the Zeiss projector beamed the celestial bodies, was composed of steel panels drilled with small holes and lined with rock cork, and was entirely without echoes or acoustic reverberation, giving the effect to the audience of being detached from the exterior world, floating in space. The effect was similar to the stillness of a great cathedral: meditative, almost spiritual. Below the Theatre of the Sky was a circular Hall of the Sun, also known as the Copernican Hall, where visitors could view scale models of the planets, including the Earth – a deliberate recreation of the set-up in Oskar von Miller's Deutsches Museum in Munich, where the projection space is balanced by the room containing models of the planets, thus offering an understanding of the solar system as both luminous experience and as physical model. On the floor at the centre of this hall was

Hayden Planetarium, Copernican Hall.

a large ceramic reproduction of the face of the Aztec Sun king, Tonatiuh, copied from an original in Mexico City – non-Christian cults lie never too far away in U.S. planetariums. The entrance hall was illuminated by lights shaped as the planet Saturn, complete with glowing rings.

At Christmas time in the 1950s the Hayden Planetarium would put on spectacular shows, with demonstrations of the skies at the time of the Nativity, dancing figures representing the celestial bodies and Christmas music – performances that rivalled Radio City's Rockettes in popularity. These shows were part of a move to add popular entertainment to drier scientific explanations.

The popular fascination with the mix of science and celestial drama was already celebrated in a Hayden publicity flier from 1945:

> As act follows act in the Drama of the Skies, you hear a human voice describe the stellar actors and the thrilling plot. Unseen in the dark, with a control board before him, stands a lecturer who has the universe at his fingertips. The control board seems the common bridge of a time-and-space ship, and the lecturer the pilot.

A few years later, in the mid-1950s, James Blish's sci-fi novel series *Cities in Flight* would feature the mayor of New York at the controls as the whole city takes off for a better life on another planet. For the moment, however, the Hayden remained in position.

In 1969, after the Apollo 11 Moon landing, the planetarium featured shows with hostesses dressed in costumes derived from those worn in Stanley Kubrick's recent *2001: A Space Odyssey* (1968), and packs of space food distributed to visitors. The Muppets performed in a Hayden show called *Wonderful Sky* and the *Star Wars* robots introduced *Robots in Space*. Like the Moscow Planetarium, the Hayden welcomed visits by celebrity astronauts, such as Edgar

Mitchell and Buzz Aldrin, linking the building to the achievements of the U.S. space programme.

Though the planetarium was housed in a conservative 1930s building, the need to attract an audience meant it had to keep up with the latest fashions in the heavens. The Hayden Planetarium repeatedly updated its Zeiss projectors, always keeping to the most recent model with the latest special effects. In 1994, however, it was decided that the Hayden Planetarium was excessively old-fashioned and unable to properly demonstrate contemporary astronomy. The building was demolished and replaced by the spectacular Rose Center (of which more later). The old shows may have lacked the special effects of today, but they suited the spirit of the period, and were perhaps more subtle and meditative than contemporary space extravaganzas.

The planetariums of the West Coast were rather different in spirit from those of the East. In the same year as the opening of the Hayden, a magnificent observatory and planetarium welcomed the public on the other side of the continent, in Los Angeles, putting all other rivals in the shade. This building was funded by the considerable fortune amassed by the mining tycoon and real estate dealer Griffith Jenkins Griffith. Griffith was a self-made man, originally from Wales, with a self-assigned rank of Colonel and an assertive personality. He donated a large area of land at the easterly end of the Santa Monica Mountains, to become Griffith Park. In 1903 he shot, but did not kill, his wife, under the delusion that she was plotting against him with the Pope. Griffith served two years in prison for the crime, but, possibly as a result of seeing the error of his ways, he stipulated in his will that an observatory, together with a cinema dedicated to astronomy, was to be built on the slopes of Mount Hollywood. Once, a wealthy repentant sinner might have built a chapel to atone for his sins; now another form of heavenly remembrance was deemed appropriate. Griffith had become fascinated by studying the stars through a large-scale telescope and was convinced that if other human beings were able to appreciate this view, their lives would be changed for the better.

The Griffith Observatory, together with its accompanying planetarium, which now replaced the cinema idea, was completed in 1935. The architect was John C. Austin, designer of a variety of well-known LA landmarks including the neoclassical Masonic Temple on Hollywood Boulevard, the neo-Moorish Shrine Auditorium for the masonic sect the Shriners, the neo-Hispanic Monrovia High School and the monumental neoclassical 24-storey Los Angeles City Hall. Austin tended to think big and historical, and the Griffith Observatory, sitting on the Mount Hollywood ridge and proudly displaying its grand Art Deco facades and large planetarium dome, is a true palace, in full Hollywood style, dedicated to astronomy. It is approached from the north by a winding roadway, Western Canyon Road, leading to a formal garden and the grand facade, while to the south the land falls away steeply, the great drum supporting the planetarium dome emerging from the rock. The

Griffith Observatory, Los Angeles, 1935.

location gives extraordinary views over the city, and to the HOLLYWOOD sign. This great palace featured a large planetarium, located in the great central dome and with the latest Zeiss projector; an observatory with a Zeiss refracting telescope in the east dome and a solar telescope in the west; and a museum with an extensive display of astronomical objects, including a Foucault's pendulum demonstrating the rotation of the Earth, and a large-scale relief model of part of the Moon. The complex has been extended and updated many times, the planetarium re-equipped with a series of the most up-to-date Zeiss projectors, and the museum featuring a Depths of Space Hall, a Cosmic Connection exhibit, the Wilder Hall of the Eye and, in the tradition of mixing astronomy with science fiction, a Leonard Nimoy Event Horizon Theatre, fulfilling in grand style Griffith's desire to be remembered as a patron of the heavens.

The Griffith Observatory naturally found its way into Hollywood films. Part of Nicholas Ray's *Rebel Without a Cause* (1955) is set in the Griffith planetarium: James Dean and other teenagers watch an astronomical show projected by the dumbbell Zeiss projector.

> Lecturer: [standing at his lectern and pointing at the illuminated sky with his illuminated lamp] As this star approaches us, the weather will change. The great polar fields of the North and South will rot and divide, and the seas will turn warmer. The last of us search the heavens and stand amazed, for the stars will still be there, moving through their ancient rhythms. The familiar constellations that illuminate our night will seem as they have always seemed, eternal, unchanged, and little moved by the shortness of time between our planet's birth and demise.
>
> ...
>
> Jim Stark (James Dean): Boy!
> Plato (Sal Mineo): What?

Jim: I was just thinking that . . . once you've been up there you know you been some place.

Plato (Sal Mineo): Do you think the end of the world will come at nighttime?

Jim: No, at dawn.

The youths become restless and distracted. But then the drama is ramped up with a depiction of exploding stars and the violent death of the universe. The Earth is described as having no importance at all at a cosmic level – its disappearance, announces the voiceover, will simply not be noticed. The teenagers are terrified by this message, but before long exit the observatory, past the giant projector, to continue their gang feuds.

The 1950s were the decade in which the possibility of the destruction of the Earth, or at least of human life on Earth, was felt to be a distinct possibility. Under the looming threat of nuclear annihilation – whether accidental or as a deliberate act of war – the immense cosmic destructions of the past suddenly took on a much greater contemporary relevance. This climate of fear was fuelled by developments in the science, and pseudoscience, of astronomy, shown in the *Rebel Without a Cause* planetarium sequence. In 1950 Immanuel Velikovsky, a Russian Jewish psychiatrist and general scientific loose cannon, published the best-selling *Worlds in Collision*, in which he argued that Venus was a comet recently arrived in the solar system, and that catastrophic events in world history, such as the biblical flood and the parting of the Red Sea, had been caused by the erratic paths of planets. In later books he proposed that emissions from Jupiter had caused the destruction of Sodom and Gomorrah, and that Mars had influenced the collapse of the Tower of Babel. Velikovsky planned to build his own planetarium in New York, to substantiate his theories, but his claims were violently disputed by astronomers and other scientists, together with historians of the ancient world. His response was to use this extreme antagonism to put himself in the position of the misunderstood genius, like Galileo, attacked and undermined by his mediocre colleagues. The scenes in *Rebel Without a Cause* show how in the mid-1950s planetariums had already moved on from simply trying to be scientific and to present the night sky as seen from the Earth, to fulfilling a more cinematic function, delighting in the scale of the universe and its destructive potential, a theme appropriate enough to a certain California mentality.

Numerous other Hollywood films used the Griffith Observatory as a location, often with links to beings from outer space, from the sci-fi B-movie *Phantom from Space* (1953), with an invisible alien trapped on the upper platform of the telescope, to *The Rocketeer* (1974), with superheroes fighting Nazi mobsters in the air above the building, and *The Terminator* (1984) with a naked Terminator arriving from the future to the surprise of some punks still hanging around from the time of *Rebel Without a Cause*. One might expect aliens and vehicles from another dimension to arrive anywhere in California, but Observatory Crest seems particularly appropriate for such sightings. More recently the original interior of the Griffith Planetarium, as featured in *Rebel Without a Cause*, is evoked in the film *La La Land* (2016), in which the characters played by Emma Stone and Ryan Gosling somehow obtain access to the planetarium at night, wander past the fizzling Tesla Coil and the Foucault's pendulum and then rise up towards the dome of the projection hall to dance in a night sky projected by the Zeiss dumbbell projector – star theatre of the 1930s becomes musical dance theatre of the current age.

Not all American planetariums were funded by wealthy philanthropists; there was also a strong current in the U.S. of build-it-yourself astronomy. In 1934 the brothers Frank and John Korkosz constructed the first non-Zeiss stellar projector for the

Rosicrucian Planetarium, San Jose.

Seymour Planetarium in Springfield, Massachusetts, a remarkable home device and triumph for American DIY technology, with 41 individual projectors within a star ball, but without any moving elements showing the Sun, planets or comets. In true DIY style, John Korkosz had started out as a child inspired by the appearance of Halley's Comet in 1910, and had constructed a comet projector from an old explosives box. Korkosz's self-built projector in the Massachusetts planetarium would inspire others to build their own projectors, including Armand Spitz, who in the 1960s would became the primary manufacturer of projectors in the U.S.

Back on the West Coast, and rather more exotic than the Korkosz brothers' planetarium, was the Rosicrucian Order planetarium, opened in 1936 in their headquarters on Naglee Avenue in San Jose. The Rosicrucians are a mystical philosophical order whose origins lie in the religious sects of late medieval Germany, but whose interests reach even further back to early Christian, Egyptian and Brahmanic beliefs. Tracing any kind of Rosicrucian set of beliefs is difficult, since there is no single order but a large number of rival subgroups with differing concerns, who compete with one another fiercely, each claiming to be the true Rosicrucians. However, all adherents share a fundamentally Gnostic position in a search for spiritual enlightenment. Rosicrucians have included the English mystics Robert Fludd and John Dee, and the German Michael Maier, all of whom had an interest in astronomy, and possibly also the noted astronomer Johannes Kepler, who interlinked pragmatic astronomy – the observation and calculation of the elliptic orbits of the planets – with more esoteric spiritual beliefs. Rosicrucian cosmology is unusual and complex, to say the least, and often explained at great length. The philosopher Max Heindel, who belonged to a rival order to the one founded in San Jose, in his *The Rosicrucian Cosmo-conception; or, Mystic Christianity* (1909), combining mystic Christian and occult theories, expounded a dense and imaginative cosmological system that included invisible worlds, seven cosmic planes and unusual theories of divine influences on evolution of the solar system. Finding points of agreement between such an individual cosmology and more standard astronomical ideas is undoubtedly not simple.

The San Jose order of the Rosicrucians, named the Ancient and Mystical Order Rosae Crucis (AMORC), was established by Harvey Spencer Lewis, originally a commercial artist from New Jersey, in 1915. Lewis became the first Imperator of the San Jose order, and created the extensive park in which the planetarium is located, financed by donations from wealthy members. In 1929 Lewis travelled with a large group of Rosicrucians to Egypt to experience at first hand the pyramids and ancient Egyptian culture, and to conduct a mass initiation ceremony in the temple of Karnak. Together with his brother Earle Lewis he is credited with the buildings in

Rosicrucian Park, though just who was responsible for the designs remains uncertain. The buildings are set out with considerable panache and with the Egyptian connection clearly in mind. The Karnak-styled Egyptian Museum, the Alchemy Museum, the Grand Temple and the Research Library are all built in full-blown Egyptian style and might easily provide sets for a reduced-scale version of the Hollywood blockbuster *Cleopatra*. The Grand Temple has a remarkable ceiling based on the night sky. Stone sphinxes and monumental pylons are scattered through the park. The Planetarium is a charming temple, a loose West Coast interpretation of the neo-Moorish style, celebrating the achievements of Arab astronomers. Green domes and turrets with spiky pinnacles rise above cream-coloured walls with tall, narrow windows and a grand arched doorway.

Lewis had visited Munich in the early 1930s and been inspired by the Bauersfeld planetarium show. He was impressed in particular by the direct nature of the experience of the projector lights producing the illuminated night sky, an immediate and for him mystical revelation linked not just to contemporary science but to a spiritual revelation as to the nature of the cosmos. In Lewis's view, the Rosicrucian interest in enlightenment through the mysteries of the cosmos might be effectively linked to the idea of the cosmos displayed through the means of light. Lewis had a background in audio-visual technology and had constructed, among various other ingenious devices, the Luxatone, which turned musical and other audio signals into colours and on which he performed before a large audience in the Francis Bacon Hall. Other Lewis inventions included the Cosmic Ray Coincidence Counter, a kind of Geiger counter for the detection of cosmic rays, and the Sympathetic Vibration Harp, a twelve-string harp that demonstrated harmonious vibrations. He also invented black mirrors that, so experimenters reported, induced cabbalistic visions, and which for a while sold in considerable numbers. Lewis saw no great philosophical barrier between ancient and contemporary scientific systems.

In that period, a Zeiss projector would have cost a quarter of a million dollars in the currency of the time, and so Lewis decided to construct his own. Following the lead given by the Korkosz brothers in Springfield, Lewis designed and made with the assistance of various technically minded Rosicrucian brothers his own star ball projector, much like the Zeiss Mark I originally constructed by Bauersfeld. This projected stars, but not moving planets or comets. A photo in 1936 shows Lewis standing, dwarfed by his projector, which resembles a large industrial case, surmounted by the star globe. A metal drawer in the case stands open, and Lewis is demonstrating to a female member of the order some transparent discs, of unknown purpose but possibly for acoustic accompaniment to the show or even film reels. On the front of the case is a simple set of control buttons.

Considering his early inventions, it would be no surprise if the Lewis projector was no ordinary machine but was also concerned in some way with cosmic rays or other manifestations. He stated:

> Differing from the other few planetariums in America or those in Europe, all of which are owned and controlled by scientific institutions, the Rosicrucian Planetarium will not be confined exclusively to a demonstration of the astronomical laws according to the Copernican theory. In this planetarium the old theories of ancient astronomers which guided the Egyptians will be demonstrated.

Somewhere here is a memory again of both the goddess Nut and also Schinkel's Queen of Night set for Mozart's opera *The Magic Flute*, with its combination of stagecraft and Masonic philosophy and its final chorus proclaiming, 'then is the Earth a heavenly kingdom, and mortals like the gods'. The delicate mysteries of the Rosicrucian cosmology would be revealed by this industrial box. Lewis often acted as lecturer for the shows – another sky-manager – and the shows would certainly have been fascinating and individual, but unfortunately were not recorded. The Lewis projector was replaced shortly after the war with a more conventional Spitz projector.

Lewis died in 1939. His remains lie under a pyramid of rosy-coloured stone in a location he selected himself within the Akhenaton Pavilion in Rosicrucian Park. On one side is inscribed the word LUX. His son Ralph Maxwell Lewis recorded his father's life in his biography *Cosmic Mission Fulfilled* (1966), whose title defines fully the achievements of the first Imperator. The Rosicrucian planetarium still continues to hold regular performances, narrated by Hollywood stars, such as 'Journey to the Stars', narrated by Whoopi Goldberg, and 'Cosmic Collisions', with a voiceover by Robert Redford.

As entertainment centres, as scientific demonstrations, as would-be spaceships, as theatres of the sky, as temples to ancient cults, as centres for mystic cosmology, as donations by industrial magnates or self-built operations by ingenious handymen, 1930s American planetariums reflected the full range of the society that produced them. Back in 1920s Moscow, Aleksei Gan had hoped that the planetarium would allow Soviet man to 'forge a scientific understanding of the world and free himself from both the fetishism and prejudice of savage priests and the pseudoscience of European civilization'. The San Jose Rosicrucian planetarium, with its interests in what Gan would have considered cults and superstitions, was diametrically opposed to the beliefs of Soviet scientists. The idea of the planetarium is open enough to accommodate many forms of belief.

FOUR

OUTER PATHS

The planetarium is scientific. The planetarium is theatrical. The planetarium demonstrates the nature of space. The planetarium brings the heavens down to Earth. The planetarium detaches the viewer from his or her usual environment.

There are various descriptions, from different periods, of what a visit to a planetarium was like, and what effect the show had on its individual spectators. For instance, the Paris Planetarium in La Palais de la Découverte, an interior space without any particularly noteworthy external architectural features, left a strong impression on the generations of children taken to visit it. The French writer Bernard Lancelot has written in a recent online article about a childhood visit to the Paris planetarium in the 1950s:

> The room of the Planetarium was completely round, with large blue armchairs in which one settled, raising one's head. We spoke in low tones, waiting patiently. On the circular walls were black silhouettes of the famous monuments of Paris, the Eiffel Tower, Notre-Dame and Sacré-Coeur. In the centre was a strange black apparatus formed of two large spheres, from which emerged into the darkness a luminous flux. There were various lecturers, but the one my sister and I preferred was a small weak-looking man, almost a dwarf, with long hair. He

> was not very distinguished. He entered limping, walking slowly towards the unusual apparatus at the centre of the room and pressed on various buttons. Night emerged slowly, and the stars appeared one by one on the celestial vault. The little man began to speak, as I recall without notes. He pointed out the positions of the stars each in relation to the others, so that we could recall them visually on a clear night. His voice was warm, not pedantic, simple, and it was easy to follow his explanations. He gave us the impression he understood everything. The sessions were quite short and I was often disappointed that they finished so quickly. Then there began a solemn music, always the same, to announce the end of the session and the return of the day. The stars faded slowly, one by one. The piece of music that accompanied the vanishing of the stars was a well-known overture from *Lohengrin*, by Richard Wagner. Each time I heard it, I got goose pimples and trembled with pleasure.

In Lancelot's description it is not just the show that is particularly interesting but also the atmosphere of the interior – the blue armchairs, the silhouettes of Paris, the great Zeiss machine, the peculiar lecturer. The visitor enters a building he knows will provide some particular experience; he sits facing inwards and looking uncertainly at the dumbbell projector, those cut-out silhouettes around the perimeter remind him where he is, but they are stage-like, unconvincing; then, as the lights go down, he is drawn into an illusionary space much larger than he can conceive. There is something of the magic theatre from Hermann Hesse's novel *Steppenwolf* (1927), as the hidden space within the city in which something unusual will occur, or of a club in a David Lynch movie, as one almost expects the lecture to speak in reverse. The lecturer is an important part of the show – his or her individual personality provides a balance to the mechanized projector. The planetarium is not just a

London Planetarium, interior overhead view.

dark, neutral space in which images of the sky are projected: it has its own qualities, which make it unusual and memorable. The best planetariums create their own particular atmosphere, their own sense of space, before the show has even begun.

In contrast, in Alice Munro's short story 'The Moons of Jupiter' (1978) a woman visiting her dying father in a Toronto hospital diverts herself with a trip to the nearby planetarium (actually the McLaughlin Planetarium, which opened in 1966 and sadly closed in 1995). The description provides an interesting contrast to Lancelot's strange memories of a childhood experience:

> There was some splendid, commanding music. The adults all around were shushing the children, trying to make them stop crackling their potato-chip bags. Then a man's voice, an eloquent professional voice, began to speak slowly, out of the walls. The voice reminded me a little of the way radio announcers used to introduce a piece of classical music or describe the progress of the Royal Family to Westminster Abbey on one of their royal occasions. There was a faint echo-chamber effect. The dark ceiling was filling with stars. They came out not all at once but one after another, the way stars really do come out at night, though more quickly. The Milky Way appeared, was moving closer; stars swam into brilliance and kept on going, disappearing beyond the edges of the sky-screen, or behind my head. While the flow of light continued, the voice presented the stunning facts. A few light years away, it announced, the sun appears as a bright star, and the planets are not visible. A few dozen light years-away, the sun is not visible, either, to the naked eye. And that distance – a few dozen light years – is only about a thousand part of the distance from the sun to the center of our galaxy, one galaxy, which itself contains about two hundred billion suns. And is, in turn, one of millions, perhaps billions, of galaxies. Innumerable repetitions, innumerable variations. All this rolled past my head too, like balls of lightning.

Munro catches the curious atmosphere of the typical planetarium show – the sense of detachment, the echo chamber sound effects, the pompous music, the solemn voice, the banalities and irritations of children both interested and bored, the feeling of scientific wonder set against the points where the illusion breaks down, the vastness of the figures and facts being related. It is clear that the 1970s planetarium show has moved on from the 1950s description of Lancelot; it is no longer just concerned with the solar system and a background of stars, but with the vast extent of the universe. How does the spectator balance those millions of galaxies with the sound of snacks being munched? The experience mixes small-scale and grandiose, banality and splendour. The character in Munro's story goes on to describe the planetarium to her father as 'a slightly phony temple', but then regrets this, saying, 'I had meant to be truthful, but it sounded slick and superior.' A slightly phoney temple is what the planetarium has been from the beginning, if one recalls those 1920s German planetariums resembling temples, places for both worship and entertainment, and then, going further back, the distant origins of the planetarium in the holy theatre, in rites and rituals. Slick and superior? Truthful? At the end of a planetarium show one is left both amazed by the quality of the astronomical spectacle, and with a feeling of having been taken in by a wonderful piece of theatre, by special effects, by that voice relating facts that are beyond comprehension. 'Awe,' writes Munro, 'what's that supposed to be? A fit of the shivers when you looked out the window? Once you knew what it was, you wouldn't be courting it.'

The planetarium also appears in literature not as a description of a real place, but as a metaphor for a certain kind of everyday life. The French-Russian writer Nathalie Sarraute's *Le Planétarium* (1958) concerns a set of characters who behave like imitation stars revolving slowly in the darkness among endlessly described domestic objects – chairs, doors, furnishings, carpets. The reader never quite knows what kind of universe, whether literal or metaphysical, is being described; nothing seems fixed, and people and objects seem to revolve in an uncertain space, repelling and attracting one another through some invisible force, equivalent to the invisible forces that order the solar system of the planetarium. One character comments, 'It is as though a fluid emerges from you, which acts at a distance on things and people; a docile universe, populated by amiable genii, organizes itself harmoniously around you.' Sarraute said of her novel, 'We are always for each other a star like those we see in a planetarium, diminished, reduced. So, they see each other as characters, but behind these characters that they see, that they name, there is the whole infinite world of the tropisms.' In Sarraute's novel the characters make no attempt to be real; they are more like figures from the Madame Tussauds waxworks next door to the London Planetarium who have somehow moved over and entered the projection system of the stars, becoming the celestial bodies for a simulation of a domestic universe.

The theme of the planetarium, and the flickering between mechanical and human, finds its way into Adrienne Rich's poem 'Planetarium' (1968), dedicated to Caroline Herschel, sister of the famous astronomer and discoverer of various comets and nebulae but who is generally rather written out of history in favour of her brother. The poem gives astronomy a feminist angle, criticizing the forms of monstrous women imagined in the constellations and ending with Caroline Herschel's body as a kind of cosmic receptor:

> I have been standing all my life in the
> direct path of a battery of signals
> . . .
> I am a galactic cloud so deep   so invo-
> luted that a light wave could take 15
> years to travel through me

Rich is not interested in an actual planetarium, but in the idea of the female body acting as an instrument for receiving pulsations from the skies and translating them into images – a latter-day version, perhaps, for the star-studded body of the goddess Nut. Both Saurraute's novel and Rich's poem show that the planetarium, for the most part invented, constructed and directed by men, could now be turned by women writers to other uses.

In these descriptions the planetarium show has begun to change its nature and purpose. The original Bauersfeld planetarium was based on the view of the heavens from the Earth, showing the moving planets and the fixed stars. The Earth was not included as one of the moving planets on the projector, because it was assumed that it would never be necessary to see the planet from elsewhere in space – it was always the stable position from which other planets were observed. But the potential space shown by the planetarium grew rapidly larger, from an enclosed system to an apparently infinite universe. When Lancelot visited the planetarium, space flight existed only in science fiction. By the time of Sarraute, Sputnik had orbited the Earth (1957), and four years later Yuri Gagarin made the first manned space flight (1961). In the year following Rich's poem, Neil Armstrong and Buzz Aldrin were walking on the Moon (1969), and by the time of Munro's story the joint U.S.–USSR Apollo–Soyuz manned space station (1978) had already been in orbit for a year. In 1972 the astronauts of Apollo 17, on the last manned Moon mission to date, took the celebrated photo in which the Earth is shown as a 'blue marble'. Once the human viewpoint moved away from the Earth and became either that of an astronaut in orbit around Earth or on the Moon, the traditional way of thinking about the solar system and outer space altered, which affected in turn the nature of the planetarium.

Robot probes expanded this viewpoint gained by astronauts, from the time of Luna 2 photographing the dark side of the Moon (1959), through the 1960s Mariner probes to Venus, Mars and Mercury, to the 1977 Voyagers 1 and 2 missions, launched to Jupiter, Saturn and interstellar space on a mission that still continues today. Most of these probes sent back detailed visual scans of planets, which were converted to resemble photos, creating a certain image of how things look in space. The operators of the planetarium had to consider how it might be possible to show our galaxy from any number of viewpoints, including those to which humans have never been, and also how to keep up with the fluctuating popular enthusiasm for space exploration and space films. Philosophical questions, such as how humans might relate to the vast space of the universe, and technical issues such as how we might display these spaces, continually overlap and have to be asked in the everyday environment of the planetarium. In fact it would take many years to evolve projectors which could show the stars convincingly from different locations in space, a problem that was only really resolved by the invention of the digital projector in the 1980s.

THE CONSTRUCTION OF PLANETARIUMS IN the 1960s and '70s was carried out against the background of the East–West rivalry of the space race, and the resulting popular enthusiasm for astronomy and all things beyond the new frontiers of outer space. Even in countries that were not directly involved in the space race there was considerable interest in astronomy's ability to reveal the ever-expanding realm of the cosmos. Planetariums became public theatres for returning astronauts, as celebrity astronauts and cosmonauts could be shown off to the public. These increasingly media-oriented adventures had the effect of making astronomy popular, and thus revived the fortunes of the planetarium. Both East and West constructed large numbers of planetariums in the late 1960s and '70s to inform the populace of national success in space, and to keep up popular enthusiasm for the projects. Many have little architectural quality, consisting mainly of small-scale domes within existing museums and schools. The first age of rich tycoons and revolutionaries who built wonderful buildings dedicated to the heavens was over.

The effect of these changes in how the universe was viewed filtered through slowly into the world of planetariums. For a long time the old, familiar planetarium solar system continued to rotate. After the end of the Second World War there was a pause in planetarium building, as the division of Zeiss into two sections – east in Jena and west in Oberkochen – meant that for the moment projectors were in short supply. Zeiss projectors were anyway complicated to manufacture and highly expensive. Zeiss in Jena became the supplier for the Eastern bloc, Zeiss in Oberkochen supplied the West, while both competed for the non-aligned world. The planetariums of the period followed the lines drawn out by the Cold War.

In contrast to these manufacturers of complex and expensive projectors, one rather unlikely man brought about a fundamental change in the nature of the planetarium: Armand Spitz, journalist, amateur astronomer, ingenious inventor, Quaker, democrat, Sputnik spotter and salesman of planetary real estate. 'I don't get along with mathematical equations,' stated Spitz, 'I am not much of a

Armand Spitz teaching children, 1950s.

Spitz portable planetarium.

scientist. You can call me an interpreter of science.' Spitz started out in the 1930s as a journalist and lecturer in astronomy. Working in the Philadelphia Planetarium he became convinced of the need for a simpler planetarium accessible to all – one rough and immediate. On the ceiling of his family home in Baltimore, Spitz painted images of the planets among the zodiac signs – a distant homage perhaps to Eisinga's domestic planetarium. He graduated to making hemispherical models of the Moon, over a metre in diameter, then by 1940 to creating The Pinpoint Planetarium, sold as part of an astronomy manual and made of cardboard with pierced holes in the positions of the stars – by placing a light inside, stars could be projected onto a ceiling or wall. His masterstroke occurred in 1945 when he invented the Spitz Model A portable planetarium. This device took the form of a dodecahedron constructed of flat metal sheets, again with holes drilled by hand and with a projector lamp inside. The dodecahedron could rotate, and later versions incorporated planets and comets. It was more primitive than Bauersfeld's first projector of two decades earlier, but retailed for $500 and was a great success throughout the 1950s and '60s. Now, not only rich plutocrats or civic authorities could own planetariums, but so could schools, universities, astronomy societies and military training schools – anyone who could afford the basic outlay. Even children were catered for: the Spitz Junior, an 18-centimetre black sphere mounted on a pyramidal base, with a 1950s, retro look, was intended for children and retailed at $14.95. Over a million were sold.

Spitz was a showman. He partly financed his business by selling 'Astronomical Quitclaim Deeds' – tax-free land rights on stars ($1), planets ($100–$250) and the Moon and Sun ($500). Certainly, it is not always easy to make money in the planetarium business. Later he evolved his Model A projector into a series of much more sophisticated devices, imitating the Zeiss dumbbell. These machines, produced by Spitz, Inc., had a looser, more self-assembled look compared to the very solid and Germanic Zeiss, and were used in many planetariums without the budget for a Zeiss, and are still produced today. The firm has grown and today manufactures all kinds of planetarium equipment, from digital projectors to domes and recorded shows. From homemade cardboard models of the Moon to a large-scale manufacturing business: it sounds like the American Dream. Something has been lost of the charm of those early days in the Spitz family home, the tables piled with astronomy models, maps of the night sky and half-completed machines. But Armand Spitz democratized the night sky, as his economically priced projectors were sold not only in the U.S. but across the world. 'I can only hope', he stated towards the end of his life, 'that in whatever celestial book-keeping there is I will be given indirect credit for helping along the knowledge of the heavens.'

Spitz had many rivals. Japanese manufacturers such as Goto and Minolta also began to produce projectors in the 1950s, again modelled on Zeiss, but cheaper. Planetariums became much more varied, from a simple room in a school containing a dozen people, to demountable and inflatable planetariums that can be moved around, to the hemispherical screens for shows in museums, to grand, state-financed public buildings with an audience of several hundred.

James S. McDonnell Planetarium, St Louis, 1963.

Centennial Planetarium, Calgary, 1967.

Since the number of planetariums continued to increase in the period from the early 1960s to the mid-1980s and the arrival of the digital world, one can only follow certain paths – sometimes unexpected and which appear to be the most intriguing. The large numbers of planetariums constructed in the U.S. were for the most part fairly banal as architecture. There are certainly exceptions, such as the elegant James S. McDonnell Planetarium in St Louis, Missouri, with its thin-shell hyperboloid roof designed by Gyo Obata. In Canada there are various interesting planetariums, such as the extraordinary late 1960s Centennial Planetarium in Calgary by the local architects McMillan Long & Associates, now sadly vacant, and the 1980s spaceship-like Space Science Centre in Edmonton by Douglas Cardinal Architects, which looks as though it has just landed after a long journey from an alien planet. But it is worthwhile to roam along less well-known paths and investigate the former Soviet bloc, London, India and South America, as each culture adapted the forms inherited from 1920s Germany and 1930s USA to their own needs.

The Soviet bloc began the revival of interest in large-scale planetariums shortly before the beginning of the space race. The second purpose-built Russian planetarium was opened in Stalingrad – now Volgograd – in 1954. Very different from the elegant modernist 1929 building in Moscow, it was an imposing block with a magnificent portico, sculpted frieze and pediment – a neoclassical temple constructed in the Stalinist style. The dome was surmounted by a large female statue of Peace holding an astrolabe and dove, by the distinguished sculptor Vera Ignatyevna Mukhina, who back in 1937 had provided the extraordinary figures of a male and female worker, bearing aloft a hammer and sickle, for the Soviet Pavilion in the Paris Exposition Universelle. No one was to doubt that this was a socialist building, erected by and for the workers, in contrast to the U.S. planetariums of the 1930s, financed by plutocrats. It was significant that Stalingrad was chosen by the Soviets for their second planetarium, for the city had been totally destroyed during the Second World War. The planetarium with its Zeiss projector was a belated 'gift' from the East German republic to Joseph Stalin, originally intended as part of the celebrations for his seventieth birthday, actually in 1949. In fact, the building looked back to an architectural style that had already almost disappeared. Stalin had died in 1953, before receiving his birthday present from Zeiss, and the architectural style that takes his name was effectively over by the time the Stalingrad planetarium opened. In the foyer is a large mural of the Marshal, dressed in the white uniform of an admiral of the Soviet fleet, surrounded by fluttering doves and blooming lilacs, against a background of shimmering gold. The mural was covered over with plaster in the Khrushchev years, but has recently been restored: the Secretary-General is back in full glory. On either side of the mural are busts of Konstantin Tsiolkovsky and Yuri Gagarin; upstairs there are now stained-glass windows – clearly added at a later date – with images of Soviet spacecraft, including Gagarin's

Stalingrad Planetarium, 1954.

capsule and the Mir space station. In the palatial rooms lined with marble columns are placed large-scale models of spacecraft. In this planetarium, with the first Zeiss projector produced in Jena after the war and with its heroic figures from the early Soviet period blended with later space vehicles, imagery derived from religion is adapted without problem to the cult of the Soviet exploration of space.

The USSR did not only go for grandeur. In contrast to Stalingrad, the minuscule Penza Planetarium – which opened the same year in the town of Belinsky, in southwest Russia – was in an eccentric timber classical building that also housed an observatory. This planetarium, in its remote location, apparently had a homemade projector in the tradition of the San Jose Rosicrucians, but in a rather different philosophical spirit. Until recently, a rather disturbing oversized statue in white marble of Lenin and Stalin discussing astronomy in a relaxed, conversational manner was located in the wooded garden just before the building.

The socialist enthusiasm for astronomy spread quickly to its satellite states, with planetariums often being used to assert national identity in the uneasy new boundaries of Eastern Europe. In 1955 Poland opened the Silesian Planetarium in a wooded park near the city of Katowice, with another generous gift of a projector from the East Germans. Once again there were political reasons for the location – Katowice, in the large province of Silesia, had been a German town, becoming Polish after the war. Locating the first Polish planetarium here, and linking it by name to the Polish astronomer Nicholas Copernicus – who had actually lived in Frombork on the Baltic coast – reinforced the Polish claim to the territory. The Silesian Planetarium is a sombre dome on a ring of concrete, which in turn rests on supports cantilevering out from the ground, the whole resembling a rather hefty 1950s flying saucer, or possibly a version of the planet Saturn, a little sorry for itself, emitting a certain morose charm. Various other planetariums followed in the Eastern bloc, each unexpectedly individual. Prague Planetarium (1960), on the edge of Stromovka park, is an elegant, early Renaissance-style temple in brown stone that was originally intended to be ringed with statues of twelve workers, as the contemporary substitution of proletarians for decadent symbols of the zodiac.

Czechoslovakia and Poland had a very distinctive approach to science fiction in the socialist period. For instance, the Czech space exploration film *Ikarie XB-1* (1963) combined a moral socialist society and evil capitalists with romantic space adventures and inventive special effects worthy of Stanley Kubrick's *2001: A Space Odyssey*. The work of the Polish novelist Stanisław Lem is often both subversive and melancholic – his novel *Solaris* (1961), set on a space station in orbit above an aquatic planet, was filmed in 1972 by Andrei Tarkovsky, who also brought to the screen the Russian brothers Arkady and Boris Strugatsky's novel *Roadside Picnic* as *Stalker* (1979). Such works are in no way Soviet space race propaganda, but offer an idea of space exploration that is social and psychological, for the most part lacking the North American interest in hardware.

German Zeiss stamp, 1971.

Something of the strange quality of these books and films is reflected in Eastern bloc planetariums. The planetarium in Minsk in Belarus, built in 1965 near a children's park, has a mysterious star theatre resembling a beehive; the Nikola Kopernik planetarium in Varna, Bulgaria, built in 1968, had (because the building, like many others of the time in Eastern Europe, was demolished) a silvery dome on top of a hefty but still stylish base; while Budapest's planetarium, built in 1977 on the edge of the Népliget (People's Park), is a black dome sitting on a thick concrete ring but carried out with a certain 1970s flair. These planetariums have a particular styling; they are meditative and melancholic, weightier than need be – qualities which suit the mood of the times.

In East Germany, Zeiss Jena unexpectedly found a second wind in building planetariums and rediscovered some of their panache

Cottbus Planetarium, 1974.

Wolfsburg Planetarium, 1983.

from the 1920s, when the firm had been the impulse behind the first big planetarium boom. The German socialist state was keen on astronomy – all schoolchildren had to have an hour's astronomy lesson every week, and enthusiasm for the subject was reflected in the quality of their planetariums. In addition, at Zeiss Jena the socialist state produced planetarium projectors of the highest quality; these were one of the few industrial products that were actually better than anything produced in the West, and thus could be used to gain both foreign currency and status.

The German Democratic Republic (GDR) is not usually thought of as a location for adventurous architecture, but its planetariums were often inventive and equipped with a high level of technology. The characteristically named Raumflugplanetarium Yuri Gagarin in Cottbus, built in 1973, was simple but stylish, with a standard dome sitting on a ring of concrete and glass – it is one of those buildings that looks like it might easily act as background in some delicately coloured socialist-realist film of the period.

Such East German planetariums were in clear competition with their West German rivals, such as the innovative stepped pyramid of the Stuttgart Planetarium, with an external steel structure designed by the architect Wilfried Beck-Erlang, which opened in 1977. In the late 1970s the East German state did a deal with the car firm Volkswagen in Wolfsburg, a city founded in 1938 for the express purpose of producing VW cars. Ten thousand Volkswagen Golf cars, in the unheard-of shades of Miamiblue and Malagared – named after locations unreachable to the citizens of the GDR – were exchanged for a Zeiss Jena planetarium. Zeiss Jena and the Braunschweig firm of Kersten, Martinoff & Struhk produced the design, with a three-quarter sphere projection hall that was later much imitated in other countries. This three-quarter sphere, constructed by the highly inventive engineer Ulrich Müther, who was also responsible for various other planetarium domes in East Germany, was made by spraying concrete onto a wire mesh structure and spanned almost 18 metres, with a minimal thickness of 9–15 centimetres, linking Wolfsburg to the earliest thin-shell structure dome by Bauersfeld on the Jena rooftop. The cars-for-stars deal proved a great success, and the Wolfsburg Planetarium, with a Zeiss Jena Spacemaster projector and the sphere clad in light-blue metal sheets reminiscent of car panels – Cosmosblue? – opened in 1983, between the Hans Scharoun theatre and Alvar Aalto's Cultural Centre, as another distinguished architectural acquisition for the small town. In the 1990s, the Wolfsburg Planetarium featured rock star shows with titles such as 'Heavy Metal Made Easier' and 'Queen Heaven'. Today, the venerable Zeiss Spacemaster stands in the foyer, having been replaced in 2010 by a digital Zeiss Fulldome, and the planetarium doubles as a popular marriage venue, with appropriate laser effects showing planetary conjunctions.

Planetariums have no particular political allegiances; the night sky belongs to regimes of all kinds. In 1981 an even more remarkable planetarium opened in Tripoli, Libya, in slightly dubious honour of the then dictator Colonel Muammar Gaddafi. Gaddafi, whose regime was at the time linked to the Socialist bloc, approached

Tripoli Planetarium, 1981.

the Jena branch of Zeiss for the design of what would be the first North African planetarium, on a site beside the coast. Zeiss's in-house architects produced a remarkable planetarium building, a 15-metre-diameter dome – once again constructed, and possibly also partly designed by Ulrich Müther – surrounded by vaguely Islamic arches and joined to a set of stylish folded concrete roofs, derived from the work of Félix Candela but also reminiscent of Müther's Ahornblatt restaurant in Berlin. The general effect is of North African kitsch married to East German technology; the folded concrete roofs of Berlin set among palm trees on the edge of the desert. 'I visited the planetarium when I was younger,' writes Muftah Abudajaja, co-founder of the Libya Design Cultural Centre and one of those now attempting to conserve the building in a time of political confusion and episodic civil war. 'Approaching it from the sea, I always saw it as a giant starfish. Architecture reflects the way of life of certain people in certain places,' he says. 'The planetarium has opened its doors to so many people over the years – as a cinema, a theatre, a planetarium – and these people have given it a Libyan soul. We must hold on to that.' The Tripoli planetarium survived the recent civil war, but at present remains closed.

Candela's thin-shell concrete roof structures were an inspiration for numerous planetarium domes and roofs, from St Louis to Tripoli to, as we shall see, Colombo in Sri Lanka. Since planetarium architecture is often reduced to a roof covering the hemispherical shell, Candela's elegant and dynamic Mexican churches and restaurants offered an attractive approach, combining engineering flair with architectural form. Since such shells could be constructed using basic materials and an unskilled labour force, they were attractive to societies without access to sophisticated building technology. They were also rather stylish and fun – welcome qualities in Eastern Europe.

The boldest East German planetarium was built in Berlin: the Zeiss-Grossplanetarium in the Prenzlauer Allee. Much of the inspiration for the construction of the building came from the astronomer Dieter B. Herrmann, director of the Archenhold Observatory in Treptow and compere of the highly popular astronomy television programme *AHA*. Through Herrmann's TV show, astronomy had become a popular contemporary topic, since it was easily conveyed to a mass audience. In addition, East Germany

Zeiss-Grossplanetarium, Berlin, 1987.

took great pride in supplying various cosmonauts to the USSR's space programme, including Sigmund Jähn, who took part in the Soyuz 31 mission in 1978 – the Salyut 6 space station was equipped with cameras supplied by Zeiss Jena.

The Berlin Grossplanetarium building has an interesting plan of circular forms radiating outwards, but the remarkable 30-metre-diameter three-quarter sphere is the centrepiece of the complex, once again constructed by the ever ingenious Ulrich Müther, with concrete sprayed onto a metal framework supporting wire mesh. The metal-clad sphere stands out among the nineteenth-century buildings of the then rather derelict but now highly fashionable urban area; bold and slightly surreal, it is an object that has arrived from another state and another time. The technical fitting out of the planetarium was of the highest standard, with a bright blue Zeiss Cosmorama projector on an elegant metal frame, with programmable computer steering for the individual projectors. In addition to this projector there were numerous lasers, and a highly effective acoustic system within the inner projection dome. The planetarium opened in 1987, a sign of celestial optimism in a socialist state. At its opening, an optimistic poem typical of the period was read out to the assembled socialist leadership:

A little fairground doesn't hurt the Alma Mater
A little picture show, much technology, a touch of Prater
This all leads to the Berlin Star Theatre
Let us pour the water of light
On the fruit of knowledge
Let us in this house both instruct and delight.

Berlin Planetarium, foyer picture by the Soviet artist Andrei K. Sokolov.

In the foyer was hung a large painting of distant stars glowing among impressive swirling cosmic matter in livid blues and red, with the insignia of the GDR shining out as Sun, by the Soviet

artist Andrei K. Sokolov, who worked with the first Soviet spacewalker Alexei Leonov. The picture was obtained from Sokolov by Herrmann himself, by devious diplomatic means. The image suggests a vast expanse of space and is celebratory but also slightly foreboding, as though there may be more awaiting in the skies than first anticipated. The East German socialist state, though its glowing symbol dominated the skies, had a period of only two orbits of the Earth around the Sun to run.

For a period in the 1990s, along with many buildings constructed in the GDR, the Grossplanetarium was threatened with demolition. But in 2016 it was completely renovated. It stands today on the edge of the extensive Ernst-Thälmann-Park, with its late socialist graphics and large-scale monument to the first East German leader – signs of a confused and now defunct state at least partially redeemed by its dedication to a search for illumination in the skies, rather than on Earth.

THE EASTERN BLOC LED the way in Europe as regards architecturally interesting planetariums, but what of England, homeland of John Flamsteed, Edmond Halley, William Pearson and many other distinguished astronomers? In contrast to most European countries, the British public had little apparent interest in planetariums until the late 1950s. The available planetarium technology was considered excessively German; surely the British could invent their own projector? However eventually national pride gave way to pragmatism and the new planetarium ordered a state-of-the-art Zeiss projector.

The London Planetarium must be the only planetarium in the world linked to a waxworks. The connection is certainly rather arbitrary – the two very different money-making activities happened to be run by the same entertainment company. But between the wax and the stars there are actually a surprising number of connections and a certain logic to their being placed together. The Tussauds waxworks had always been considered a curious mix of respectable and seedy, real and illusionary, morbid and life-giving. Madame Marie Tussaud herself had begun her career making wax copies of the heads of people guillotined during the

London Planetarium, exterior.

French Revolution. During the nineteenth century the waxworks attempted to become respectable – Tussaud modelled the English royal family and heroes like Admiral Nelson, and created tableaux of famous scenes from British history, including the execution of Mary, Queen of Scots and the murder of the Princes in the Tower. A cinema that previously existed on the site had been destroyed by German bombing in 1940, in a direct hit that obliterated everything except, rather curiously, the wax figures of Adolf Hitler and other leading Nazis. By the 1950s the whole idea of a waxworks had become increasingly dull and dusty, as popular entertainment became oriented towards television and cinema, and as people could see what celebrities looked like from photographs in magazines, as opposed to the often rather dubious waxwork likenesses. But with the destruction of the original cinema there was now enough space to develop a planetarium as an independent building – a bold move designed to raise the reputation of the waxworks as an institution, projecting a forward-looking scientific and educational image. The two buildings would balance one another, the planetarium recreating the stars and planets and the waxworks providing imitations of living and dead celebrities.

The London Planetarium opened in April 1958 and is credited to the architect George Watt, together with the engineers R. Travers Morgan & Partners. Watt, a Scot who had emigrated south, specialized in bomb-damaged sites. Nothing he built before or after has quite the same elegance or refinement, so it is possible he was just a networker, while some other talented architect in his employment designed the actual building. The planetarium's architecture is based on a conservative classicism, not that different in inspiration from the building in Stalingrad, but of course rather more modest and buttoned up, as suits the English way.

The most prominent part of the building is the dome, constructed of reinforced concrete and clad in copper, which is now weathered to a lustrous green. This dome sits on a raised concrete platform, in turn supported by a ring of concrete columns with special foundations to reduce vibration from the tracks of the London Underground, whose Circle, Metropolitan and Hammersmith & City lines run underneath most of the length of the Marylebone Road on which it stands. At 20 metres, the diameter of the dome was determined by the requirements of the standard planetarium projection equipment of the period, and is rather too large for its site, squeezed between a small side street and the bulk of the main Tussauds building. But seen from the Marylebone Road, the whole effect is strong and simple compared to the heavy-handed historicism of its neighbours, Baker Street underground station and Madame Tussauds itself. The building is both classical in form but, like many planetariums, also slightly absurd. It resembles a

London Planetarium, projector drawing.

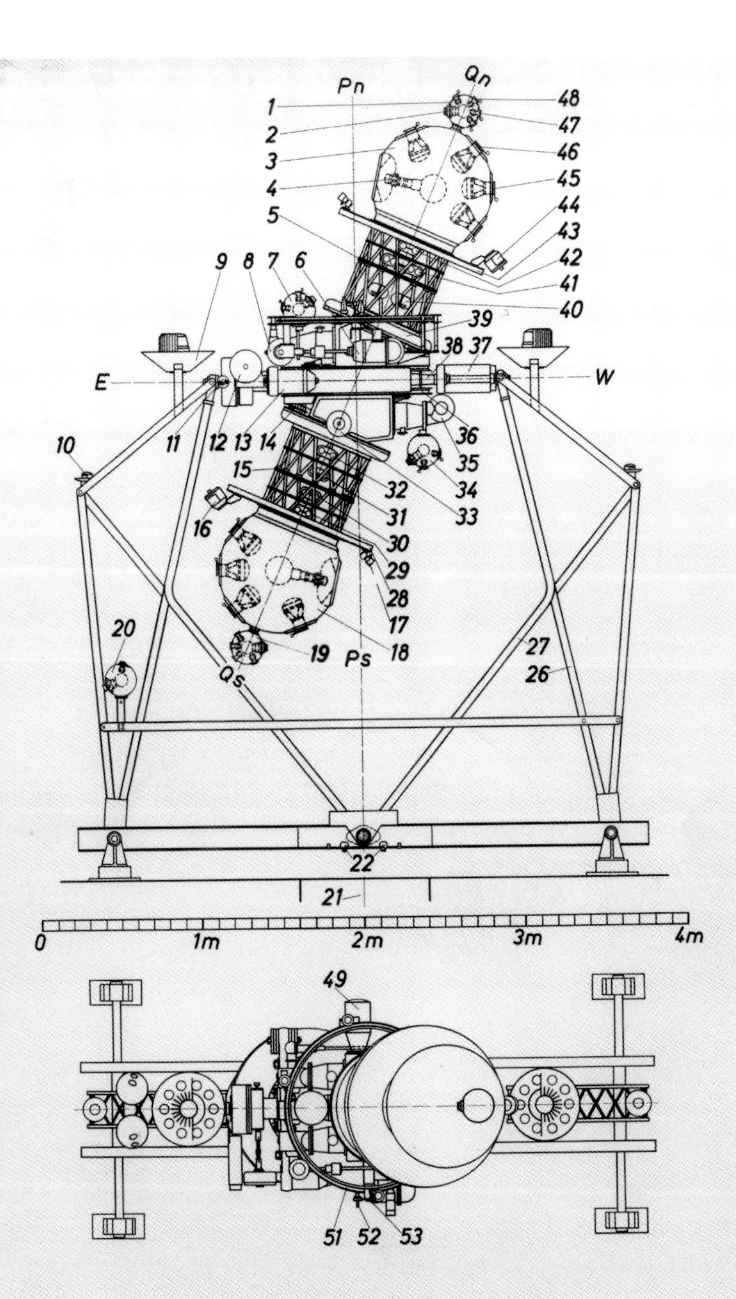

Pn
Qn
1
2
3
4
5
6
7
8
9
10
11
12
13
14
15
16
17
18
19
20
21
22
26
27
28
29
30
31
32
33
34
35
36
37
38
39
40
41
42
43
44
45
46
47
48
49
51
52
53
E
W
Ps
Qs
0
1m
2m
3m
4m

Roman temple, but also an oversized serving dish enveloped by its domed metal cover, as well as the upper hemisphere of some green planet, with the horizontal plane as its ring – an unusual mix of ancient religion, gastronomy and astronomy. Above the dome, on a post, is the smaller, white planet of its rooftop, the celestial familiar whose temporary disappearance was mentioned at the beginning of this investigation.

Today the building is much altered. At street level, an illuminated sign used to run around the base of the dome, its letters spelling out in clear typography THE LONDON PLANETARIUM. A thin horizontal plane projected out from the dome, acting as a kind of canopy, its underside inscribed with the twelve figures of the zodiac. Entry to the planetarium was directly from the street into a foyer below the domed space. This lobby was designed in the stripped-down classical-modernist style of the period, with vertical fins faced in Perlato and Napoleon marble, a mosaic floor and a rather elegant freestanding ticket kiosk. The ceiling was originally painted sky-blue, and the day-lit foyer, horizontal and open, prepared the visitor for entry to the night sky auditorium above, which remained dark, vertically oriented and enclosed. A grand stairway at the rear of the foyer linked the two levels.

Around the horizon of the auditorium ran a frieze showing the skyline of London, returning this interior, with its cosmic pretensions, firmly back to its English site. This feeling of Englishness was reinforced by the plummy voiceover by the original director of the London Planetarium, John Ebdon, a self-taught astronomer, former actor and future radio commentator, who provided the live commentary and – in contrast to the 1950s lecturer in the Paris Planetarium mentioned by Bernard Lancelot – gave to the grand, celestial movements a comfortable, slightly whimsical feel. However, the building suggests more than it reveals, and the homely feel is deceptive. Somewhere here are unconscious echoes of the most ancient notions of cosmology: the Earth as a flat disc, unmoving, raised up on columns, which in turn stand on some uncertain firmament surrounded by an endless sea of other matter, while the sky is a slowly rotating projection of stars, finite in its dimensions, and all set in motion by some seemingly divine force.

The London Planetarium functioned as a semi-independent enterprise, run by the British Planetarium Society and with its own entrance, separate from Madame Tussauds. It looked down on its neighbour somewhat, considering the scientific demonstration of the workings of the cosmos to be a higher calling to making images of dubious historical figures of criminals and celebrities. In fact, both were in the business of simulation, creating imitations

London Planetarium, publicity shot.

of the natural world, but using a different scale and one working with beams of light, the other with wax. Neither was completely convincing – the waxworks always have a sickly, dead-eyed look, as though inhabiting some sunless afterlife, while the stars and planets are of course merely ingenious projections. As time went by, however, the Tussauds waxworks began to infiltrate the planetarium, as though Nathalie Sarraute's novel, published in the same year as the London Planetarium opened, was coming to life. Wax figures of attendants were placed in the foyer to playfully confuse the visitor. These were followed by figures of famous contemporary astronomers such as Patrick Moore, the eccentric presenter of the late evening BBC television series *The Sky at Night*, which had launched in 1957, the year before the opening of the planetarium. Figures of distinguished astronomers were also placed in the foyer – Copernicus, Kepler, Einstein and others – in theatrical poses. Visitors could feel at home with these scientists of the past, just as they had previously sauntered among members of the royal family, football celebrities, film stars and notorious criminals. From 1969 Neil Armstrong and Buzz Aldrin awaited in the main Madame Tussauds building, where they stood on a rather diminutive Moon, dressed in their space suits but incautiously carrying their helmets under one arm.

The London Planetarium's own potential as a location for a fictionalized cosmic disaster was spotted early by J. G. Ballard and brought out in his novel *The Drowned World* (1962), which appeared just four years after the building opened. The novel is set some time in the future, when the world has been flooded due to the icecaps and permafrost melting – a latter-day Velikovskian scenario based not on errant comets but on the overheating Sun. The climate has relapsed into a kind of Jurassic state, conducive to oversize lizards and insects. The few remaining human inhabitants live in the upper storeys of luxury hotels and other tall buildings protruding above the swamp-like waters. The dome of the planetarium can be spotted several metres below the surface of the waters. The narrator, Dr Robert Kerans, decides to enter the building, and descends in a diving suit to street level. The building looms up, covered with molluscs, bivalves and the fronds of marine flora. Kerans finds all the pieces of the interior in place – the ticket kiosk, the staircase to the auditorium, the manager's booth.

> The dark vault with its blurred walls cloaked with silt rose up above him like a huge velvet-upholstered womb in a surrealist nightmare. The black opaque water seemed to hang in solid vertical curtains, screening the dais in the centre of the auditorium as if hiding the ultimate sanctum of its depths. For some reason the womb-like image of the chamber was reinforced rather than diminished by the circular rows of seats, and Kerans heard the thudding in his ears, uncertain whether he was listening to the dim subliminal requiem of his dreams . . . The deep cradle of silt carried him gently like an immense placenta . . . Far above him, as his consciousness faded, he could see the ancient nebulae and galaxies shining through the uterine night, but eventually even their light was dimmed and he was only aware of the faint glimmer of identity within the deepest recesses of his mind.

Ballard's description of the interior of the flooded planetarium is precise and clearly written from the memory of an actual visit or series of visits, and offers a good contrast to the descriptions quoted earlier. The planetarium scenes form the centrepiece to the book, mixing many of Ballard's continuing concerns – urban decay, the subconscious, flooding, ecstatic states gained through extreme experiences of the cosmos. The planetarium is not just a scientific instrument; it is the place where one goes to experience a rhapsodic state before the wonders of the star systems, a state that pushes one back into a memory of pre-birth. The building is in Ballard's description both tomb, filled with dead objects, and womb, nurturing the next stage of human existence.

As yet, the Ballard floods have not placed London underwater; the London Planetarium is closed for other reasons. Predictably, it was the waxwork celebrities that would eventually defeat the real celestial stars. In 1995 the planetarium sold its soul, replacing the now outmoded mechanical Zeiss projector with a smart new Evans & Sunderland Digistar 2, able to offer all the advantages of digital projection, including 3D journeys into space, even if the resolution on-screen was lower than that of the old Zeiss. But the heart of the show was gone, and the Zeiss machine was shipped off to the Alton Towers amusement park, where dismantled elements of the once magnificent technology today lie abandoned among the rides. The congested urban site, unlike that of the Adler in Chicago or the Hayden in New York City, offered no space for expansion. In the 1990s, Madame Tussauds, under new management, speeded up the astronomical shows from 45 minutes to ten, in order to increase visitor flow, resulting in extensive motion sickness among younger viewers, with predictable results. The foyer on the ground floor was taken up with a ride through the past of London, with waxworks of famous citizens such as Henry VIII and Queen Elizabeth II nodding in jerky, automated movements to passers-by. Finally, the Faustian pact reached its inevitable conclusion, and in 2006 the astronomy shows closed. The projection hall is currently used as a 4D cinema showing superhero movies.

A replacement for the Marylebone Road planetarium, the Peter Harrison Planetarium, was built in 2007 as part of the Greenwich Observatory in southeast London. It is an underground venue with an audience capacity of only 120, seated facing one direction under an 11-metre dome, onto which the sky is beamed with a digital projector. At ground level the position of the subterranean projection room is marked by a large grey cone, its southern side sloped at an angle of 51.5 degrees, and with a groove that always aligns with the Northern Celestial pole, near Polaris. The top of the cone is cut by a flat surface, at the same angle as the surface of the ground at the North Pole. Since not all of this is obvious to visitors, the planetarium possesses some aspects of a hermeneutical object, with its own astronomical secrets. The Greenwich Planetarium is a fine piece of architecture, but it is no substitute for the centrally located London Planetarium, which had been a landmark of the city, an important venue for schoolchildren and a sign for a half a century of the population's interest in astronomy. The illuminated planet on the roof has returned, but the spirit of the real stars has departed.

FOR TRULY INNOVATIVE PLANETARIUMS in the Cold War years, one must look beyond the division of the world into East and West, to the non-aligned countries. Indian planetariums in the 1960s are distinctive in that architectural typologies belonging to the local culture were adopted to house a European invention. India has a long astronomical tradition reading back to 1200 BC, with astronomy being studied as one of the Vedas, the oldest Sanskrit texts. In early medieval times India produced the most advanced astronomical observations, with the fifth-century *Aryabhatiya* offering a complex cosmology with methods for calculating the movements of the planets and other celestial bodies. India has its own history of linking architecture and astronomy. The extraordinary eighteenth-century collection of large-scale astronomical instruments at Jaipur known as the Jantar Mantar ('calculating instrument') was constructed by the Rajput king Sawai Jai Singh in 1724–38, and showed an alternative method of studying the skies. Just as European scientists were constructing ever finer instruments on a reduced scale, with complex lenses and machinery, the Jantar Mantar went in the reverse direction to build devices at a large scale. The nineteen instruments, which among other activities track stars, measure the passing of time and observe the movement of the planets, cross over between the scales of architecture and sculpture with forms that resemble flights of stairs, hemispherical cavities and circular colonnades. The Argentinian writer Julio Cortázar photographed the constructions in 1968 and wrote in his lyrical *Prosa del Observatorio* that these constructions were really a defiance of the tyranny of astronomy:

> Every scientific manual and tourist guide describes them as apparatuses designed for the observation of the stars, exact and clear and of marble; but also there is the image of the world as sensed by Jai Singh; these devices were not only constructed to measure the course of stars, to domesticate such insolent distances; Jai Singh, guerrilla of the absolute against astronomical fate, was dreaming of something else . . . a grand response of a total image against the tyranny of the planets and their conjunctions and ascensions; Jai Singh, diminutive sultan of a declining kingdom, channelled the astral light.

An image against the tyranny of the planets – or against the control of astronomy purely by Western science? Cortázar seems to be suggesting that the apparent purpose, astronomical observation, was not the only reason for these constructions, and that really their origin lies in some other hidden rationale. Indian planetariums follow on, in their own way, from the Jai Singh observatory. All planetarium buildings have a functional purpose related to science, but their architecture often suggests they are also linked to religion, early theories of the cosmos and lost cultures. In addition, Indian planetariums have a peculiar link to those in the West, since

Birla Planetarium, Kolkata, 1963.

often they are fitted with analogue projectors that were previously used in Western planetariums, considered outdated but then renovated and put back to use in India, where individual mechanical ingenuity has continued long after it faded in Europe and the U.S. The projector technology is inherited, but sometimes applied to slightly different purposes than originally intended.

The first Indian planetarium opened in Pune in 1953 with a second-hand Spitz projector from Philadelphia and a small dome for 100 visitors. The local ability to keep alive ancient machinery meant that this projector, which has a better quality of resolution than many digital projectors, has been repeatedly repaired and is still functioning. Other Indian planetariums are linked to political or business figures, who have provided funding. The M. P. Birla Planetarium in Kolkata, built in 1963 and named after the founder

Sardar Patel Planetarium, Baroda, 1976.

of a large industrial concern, has an entrance like a cinema, with Birla's name in large letters in three languages, a plan of concentric rings derived from the Adler Planetarium in Chicago, and a dome loosely based on the Buddhist stupa at Sanchi. It successfully merges these separate traditions. The Nehru Planetarium (1965) in Porbandar, Gujarat, birthplace of Mahatma Gandhi, is reminiscent of the Raj, with a long, low facade on which is placed an unusual dome surmounted by another, smaller dome.

The most architecturally innovative in creating an overlap between contemporary engineering and traditional religious architecture is the planetarium in Colombo, Sri Lanka, which opened in 1965 as part of an industrial exhibition and was designed by the remarkable engineer and expert in precast concrete A.N.S. Kulasinghe. Kulasinghe combined his pragmatic expertise in concrete with a guiding belief in the principles of Buddhism. Among his other projects are the 1956 Sambodhi Chaithya, at the entrance to Colombo Harbour, a large stupa mounted on a platform of crossed arches, creating an extraordinary mix of traditional religious form and contemporary engineering; and the Kotmale Mahaweli Maha Seya, another stupa in the form of a 60-metre diameter ultra-thin dome, out in the countryside by the Kotmale Reservoir, begun in 1983 but still incomplete. Either of these, with their extraordinary domes, would make excellent, if rather inaccessible, planetariums.

The Colombo planetarium is based on the form of a lotus flower, and is constructed from a series of prefabricated folded-plate concrete elements manufactured in Malaysia and shipped over to Sri Lanka. At the pinnacle the structure launches out into a crown of pointed spikes; at the base the structure folds outwards and upwards, both to provide structural strength to the folded elements

Sri Lanka Planetarium, Colombo.

and to offer an entrance into the planetarium. Once again the building has a resemblance to certain works by Félix Candela; it is like the sculpture for the Plaza de los Abanicos in Mexico City, but now vertically inverted. Zeiss Jena, on another of those generous GDR deals to developing countries – in the early 1970s, Sri Lanka temporarily became a democratic socialist republic linked to the USSR – provided the AG Star projector, so the building has East German links. However, Kulasinghe was clearly his own man and needed no assistance from anyone. The structure is an intriguing blend of engineering know-how and form derived from religious tradition. It continues to stand today, beside a small lake. Within, the projection space is still equipped with wooden chairs, each with a specially designed headrest for lying back and looking up at the artificial sky. On clear nights the area outside is used by large groups of children, clad in white uniforms, who queue up to observe the sky through telescopes, while Kulasinghe's building is gently illuminated and appears as a botanically inspired alien vehicle, uncertain as to whether to ascend or descend.

IN LATIN AMERICA, MATTERS ASTRONOMICAL have their own quality. The first large-scale planetariums to be built after the Second World War were in South America, they share the individuality of the Indian buildings. Montevideo Planetarium opened in 1955, constructed on the initiative of the city's mayor, the amateur astronomer Germán Barbato, and bearing his name. Montevideo may seem an unlikely location for the first Latin American planetarium, but in contrast to the Eastern bloc, where the impetus came from the state, or the U.S., where they were linked to education and the military, many Latin American planetariums originated in the enthusiasm of amateur and professional astronomers. Perhaps this accounts for their freshness and freedom from architectural tradition, linked to a desire of municipal governments to provide

Aristóteles Orsini Planetarium, São Paulo, 1957.

scientific demonstrations to the people. The Montevideo building is a simple domed building on the model of the 1920s planetarium in the Jena Prinzessinnengarten, with the first ever Spitz projector, since nothing was yet available from Zeiss at the time. This Spitz projector is still in use today. Public enthusiasm for the new building appears to have been considerable; its first director Nigel Wolf wrote, in what might be a typically South American concern for excessive popular unrest due to that arrival of a new cultural institution leading to military control: 'The excitement at the opening of the planetarium in Montevideo is extreme . . . but certainly not to the point of needing squads of riot police to control the crowd demanding admission.'

Other South American countries followed the Montevideo initiative. In 1957 a planetarium opened in São Paulo, located in Ibirapuera Park, which was landscaped by Otávio Augusto Teixeira

Mendes and with various buildings designed by Oscar Niemeyer. The park was laid out as a large green space for public use, with a popular mix of cultural buildings and football pitches, and was seen as an overt display of the arrival of modernity in São Paulo. The planetarium, together with the nearby astrophysics school, was a deliberate display of the modernity of São Paulo's astronomers and scientists. The architects for the planetarium were Roberto Goulart Tibau, Eduardo Corona and Antonio Carlos Pitombo, who followed the Brazilian modernist architectural line established by Niemeyer. The contrast between the São Paulo building and the one in Stalingrad, finished only three years earlier, or the London Planetarium of a year later, could hardly be greater. The building is light, simple, contemporary, fun – really the second truly modernist planetarium to be built, after the Moscow building of 1929. The building is partly sunken, with an exhibition room at lower level and the projection room above. The projection hall is covered with a hemispherical concrete shell covered in aluminium sheeting, with a bright yellow rim, and a lightweight aluminium canopy projects out over the entrance, giving the whole something of the appearance of a flying saucer that has landed among the greenery, but, in a another mixed analogy, also of a jaunty baseball cap. Entrance is through a curved ramp, which descends into the darkness before emerging again into the light, a feature that would later be repeated in Niemeyer's project for the Cathedral of Brasília. Generations of São Paulo schoolchildren visited the building and watched the show, which concentrated on the sky as seen from the city. In a way similar to the planetariums in Paris and London, the building has become associated with the typical experience of childhood, to be later recalled by the majority of São Paulo adults.

On a par with the São Paulo building is the Buenos Aires Galileo Galilei Planetarium. A group of Argentinian astronomers had visited the planetarium in Jena in 1927 – a photograph shows them

Galileo Galilei Planetarium, Buenos Aires, 1968.

sitting in the wooden chairs, dwarfed by the size of the Zeiss dumb-bell projector. They proposed a planetarium for Buenos Aires, but without success. Over three decades later, in 1959, a second group of astronomers attached to the now socialist city government finally managed to commission the planetarium, as part of a movement to open out the culture of the city and to make science available to all. The new building opened in 1967, in the Parque Tres de Febrero, formerly the estate of the nineteenth-century Argentinian strong-man Juan Manuel de Rosas, but now an extensive park of groves, ponds, rose gardens and a boating lake near the River Plate.

The architect was Enrique Jan, from the city's department of architecture, an unusual figure in that he only produced this one building, and spent ten years of his life bringing it into existence. A photograph obtained from his widow, Beatriz Cordon Jan, shows a square-jawed man clad in an argyle-pattern pullover, perhaps in a garden, half-obscured by a branch, peering out to his right as though not fully fixed in the photograph but drifting away to some other space. Originally, Jan wanted the dome of the new planetarium to be mounted on a single column, with a spiral staircase, in what he referred to as the style of Frank Lloyd Wright, but this first proposal lacked structural credibility. He produced a second version, keeping the column with the stair but now with three reinforced concrete legs. The exterior dome is of 8-millimetre-thick prefabricated concrete panels and is 21 metres in diameter; it contains an interior aluminium hemispherical screen for the Zeiss Mark IV projector, the first to be produced by the new western branch of the firm in Oberkochen. Like the planetarium in São Paulo, the building is partly sunken into the ground, but the main dome and a circulating promenade are supported on the three legs, which lift the main bulk off the ground and give the effect of a planet – or again a flying saucer, or even, due to the pattern on the panels of the dome, a straw hat – hovering over the park.

A planet, an alien spaceship, a hat – what was Jan seeking to achieve? At the time the architect described his project, ambitiously but also rather obscurely, as 'a suggestion of what has happened since the first elementary particle of matter to the cosmic development in which we are immersed'. The planetarium interprets an evolutionary process, with the equilateral triangle of the base becoming in turn a hexagon, then a rhomboid, and finally reaching the perfection of the circle. Jan was both a pragmatic architect concerned with form and structure, and a mystic, who saw the triangle as a symbol of divinity, of the Holy Trinity. He was renowned for remaining silent and looking up at the sky when asked to explain his architectural decisions. In a later text, *Claves para entender el el planetario* (Keys to Understanding the Planetarium, published posthumously in 2007), Jan went further:

> At this time I had some interest in the East, in particular in the synthetic capacity of art and in written language, in ideograms. The information is there, for those who can read it. The planaterium is an architectural ideogram . . . the visitor arrives via a bridge made of triangles, crossing from outside to the inside of the building. The triangle is the first elementary geometric figure capable of enclosing an area in two dimensions . . . the triangle leaps from two-dimensional to three-dimensional space, forming two inverted tetrahedrons, one supports the base on the ground and raises its peak to heaven, and the other comes down from heaven to Earth, interpenetrating the first.

Jan adds other ideas, relating to the planetarium and the flow of time: 'The nature of time is circular, it is perceived in the changing seasons, in the cycles of birth, life and death . . . the circular gallery surrounding the dome which is raised up seeks to convey this idea.' And then, a comparison to the human form:

> The central axis of the planetarium is a hydraulic lift which connects the deep to the high, as the backbone of a human being joins the sacrum (a bone strangely called sacred) and the cranial vault, inside which take place virtual representations of the perceptual world around us.

Clearly, if the Buenos Aires planetarium is an ideogram it begins to assemble numerous meanings – not only as planet, spaceship and hat but now as the purveyor of sacred geometry, time machine, and metaphor for the human form and our capacity to summon up virtual images.

Little of this is visible from the outside; one has to search in the elegant plans and sections for traces of the sacred geometry. It doesn't matter – as Jan says, the ideogram is there for those who wish to interpret it – for the Galileo Galilei Planetarium is wonderful anyway: the interior is lightweight and spacious, and the surface of the exterior dome is illuminated at night with pin-points of light, becoming yet another metaphor for the starry sky itself.

The building was used for various other functions, linked to astronomy. In 1979 it was a performance space for the *Cinco Preludios Cósmicos*, a musical piece by the accordion player Alejandro Barletta, bringing back the notion of planetary music and harmony of the spheres once explored by Johannes Kepler. That same year, the exterior of the planetarium displayed large-scale models of the Czech-Argentinian sculptor Gyula Kosice's *Hydrospatial City*, a series of proposals for planet-like cities in space. Kosice described these cities in the spirit of the times as 'habitats full of unclassifiable worlds, places for intermittent vacations, and poly-dimensional places where one could be dead and alive, hunt from prehistoric auroras, or direct satellites by remote control aboard a Kosicean space craft'. A photomontage of the period shows a line of baffled schoolchildren passing under various translucent circular forms which mimic the great spherical dome of the building.

There is an urban myth attached to the Buenos Aires planetarium recounted in a mischievous manner by the Buenos Aires writer and architect Gustavo Nielsen. According to Nielsen, Enrique Jan was not only mystically inclined but much enamoured of Ray Bradbury's collection of sci-fi stories *The Martian Chronicles*. In 1959 Jan was given a present of a first edition in English, which had been published in 1947. Bradbury had proclaimed that first edition copies of his book had special protective powers – no doubt aiding the sales of his book. Nielson relates, 'Jan divided the miraculous book into three parts, and placed each in a metal box onto which he soldered a lid. Then he concealed them in the reinforced concrete . . . Several workers saw him do this. This didn't matter to Jan. His building was "protected".' In 1997 Bradbury visited the planetarium while in Buenos Aires for a literary conference; a photograph records his visit. According to Nielsen, Bradbury attempted to find the three fragments of the book and became lost in the space between the exterior dome and the hemispherical screen. The story has now taken on its own life and is included – as a true account – in a leaflet given out by the planetarium. And indeed, why not – for so much remains mysterious and open to interpretation in this building that one layer of fiction simply adds to the mix.

Buenos Aires Planetarium collage with Hydroespacial.

Morro Solar Planetarium, Lima, 1960.

The Buenos Aires planetarium still hovers over the park, having survived decades of turbulent politics, regime changes and the horrors of the Dirty War, which brought the scientific culture associated with the building to a standstill. It remains forever both modern in feel and hermetic. A large meteorite on the ground before the building, and a small Moon rock in a case in the circular gallery, brought back from the Apollo 11 mission, offer other geological-celestial delights. At night the building is surrounded by an unusual combination of amateur astronomers with their telescopes, benefiting from the clarity of the atmosphere to study the heavens, and numbers of transvestite prostitutes and their clients, with more terrestrial inclinations.

And finally to the most evocatively sited Latin American planetarium, the Morro Solar, which opened in 1960 above the Peruvian capital, Lima. This building was another constructed by enthusiasts: the seven members of the local astronomy society, consisting of five engineers, a priest and a doctor, and designed by the engineer astronomers José Castro Mendívil and Victor Estremadoyro. It stands on the red earth of a ridge overlooking the city and the sea, between a small observatory and a half-ruined church, in a location that has religious significance for the pre-Inca population, the Ichmas. Out here one might well be aware of some of those immaterial waves suggested by Adrienne Rich. Above the planetarium are a diverse selection of monuments: an illuminated cross built of pylons, dedicated to Pope John Paul II; a 37-metre-tall statue known as the Christ of the Pacific; various military monuments; and on the crest, large numbers of transmitter antennae. This location was chosen by the astronomical society in the late 1940s on account of the clear light, which is ideal for astronomical observations, and as in the case of the Los Angeles Griffith Observatory, the planetarium was simply built alongside the existing observatory. The dome is covered in aluminium, with a projecting horizontal rim, and sits on a plain brick wall; it is a reduction of the idea of the planetarium to the bare minimum. The perimeter of the interior of the dome is painted with a view of the main buildings of Lima – a variation on the cut-outs which often feature in planetariums of the period. Even the star ball projector was designed and constructed by the engineer Castro Mendívil.

Who would visit a planetarium up here, so far above the city? A photograph from the 1960s shows the building surrounded by a number of large automobiles, all parked on the ridge, facing the sea. It remains the oldest planetarium in the world that is run by enthusiasts, and must be popular, since it was recently renovated by a group of digitally minded astronomers, and the ageing star ball was replaced with a new digital projector. Three times a week, shows are put on for 200 spectators. The better-known writers of Peru have remained uninterested, but a recent article in *Perú.21* reports: 'This morning, the place is packed and surrounded by children from three different schools. They have come to see the show in which you can watch digital projections of the sky and the formation of planets and stars and constellations. They run, they point out and they jump all over the place.' Within the projection hall is a display of objects reflecting the diversity of the world outside – telescopes, rockets, aerolites, Ichma remains, fossils, military hardware. There's no great feel of urgency here. No space race. Just a simple metal dome up on a ridge.

FIVE

VISIBLE, INVISIBLE

Ah, for the simpler times of the Egyptian Sun goddess Nut. Her body, decorated with stars, covered the surface of the Earth. She swallowed the Sun in the evening and gave birth to it again in the morning – a representation of the eternal reincarnation of the human soul. Such ideas, even if essentially non-visual, could be demonstrated with a wall painting, the architecture of the pharaonic tomb matching the myth of the goddess. The vision of the night sky could be understood by anyone, even if the complexities lying behind the idea were at that time comprehensible only to the priests.

Bauersfeld's Jena planetarium had its own simplicity. It could show the system of planets moving around the Sun – not as charming as the myth of Nut, but understandable to all. The display of the star theatre mirrored what might be seen by an observer on a clear night sky; what is visible. Again, the complexities of the astronomy of the time were not explained; they lay already beyond the ordinary observer, reaching much further than the simple view of the actual night sky. The architecture of the planetarium dome was appropriate to the notion of a solar system with definite boundaries, mirroring the actual view from Earth of the night sky. For a brief period, the exterior and interior architecture of a planetarium could be part of the same system.

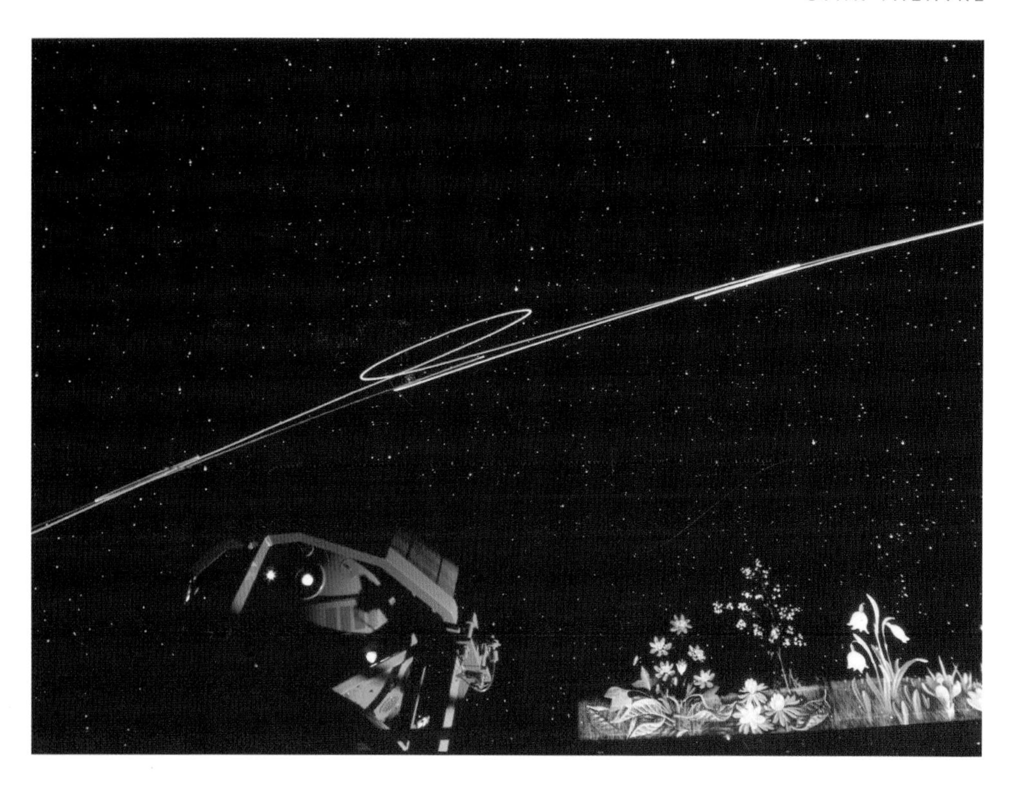

Zeiss projection of the orbits of Mars and Jupiter, 1996.

Today, however, we need to consider numerous cosmological phenomena, which are harder to match to any finite architectural volume, since these phenomena are themselves often without clear form or do not even exist within our optical field. The list runs on with a certain astronomical lyricism: black holes, brown stars, quasars, pulsars, cosmic rays, neutrinos of various flavours, weakly interacting massive particles (WIMPS), uncertainty principles, perturbation theory, wormholes, white holes, axions, dark valleys, dark fluid, infinite seas, alien communications, multiverses and many others that as yet remain notional in a field in which scientific speculation roams at will. Astronomy and its fellow disciplines such as cosmology and astrophysics have become the wonderful location for a rich diversity of possible phenomena, equivalent to those figures of extraordinary animals and humans put forward by medieval astronomers to fill the vacant spaces of heavens, or the classes of angels and other celestial beings once hypothesized by Renaissance cosmologists. Many of these phenomena emerge out of the theoretical propositions of contemporary science, but few can be observed with the naked eye from the surface of the Earth, most requiring highly sophisticated equipment to be detected, investigated and interpreted. We can see almost nothing, both because such phenomena are too distant and because many are in any case invisible. The visible and the purely material have gradually been reduced to a minor role in a cosmos now considered to be essentially non-visual and immaterial. Astronomers today speculate – but cannot really confirm, for theories and counter-theories evolve rapidly – that dark matter and dark energy makes up well over 90 per cent of the universe. We humans are thus not even made of the same stuff as most of the universe, and are relegated to a bit part in a much larger story. Elements of this speculation drift back into poetics, and to ancient myths that seemed long abandoned. In his book *Nothing* (2009) the astrophysicist Frank Close compares the current search to distinguish what exists from what does not – or may not – exist to the ancient sutras of the Rig Veda: 'The non-existent was not; the existent was not / Darkness was hidden by darkness / That which became was enveloped by The Void.' The precision of astronomy approaches the vagueness of mystical poetics. A planetarium demonstrating the principles of nothingness would be an interesting proposition.

The dramatic theatre, in its more experimental fringes, has long been interested in the idea of a theatre of nothingness and of invisibility. Invisibility does not imply that there is nothing; merely that certain things cannot be perceived, and that what lies beyond our senses only feels like nothing. Peter Brook asked in *The Empty Space*: 'Can the invisible be made visible through the performer's presence?' wondering how the actor could reveal his or her aspects of an interior life otherwise concealed. The Polish theatre director Tadeusz Kantor proposed the cry of 'further on, nothing!', implying – not entirely clearly, and with a certain mournful pleasure – that there was only a vast emptiness to anticipate. Kantor wrote in his notebooks of the 1940s of

> Space,
> which does not have an exit or boundary;
> which is receding, disappearing,
> or approaching omnidirectionally with changing velocity;
> it is dispersed in all directions: to the sides, to the middle;
> it ascends, caves in . . .

The controversial Austrian playwright Peter Handke wrote in his highly experimental play *Offending the Audience* (1969) – which is to be performed in darkness: 'You don't see a darkness that pretends to be another darkness. You don't see a brightness that pretends to be another brightness. You don't see any light that pretends to be another light. You don't hear any noise that pretends to be another noise' – as though theatrical performance had no need to represent anything of the exterior world, a suggestion that the planetarium has yet to consider. Theatre has become more conservative since the heady days of Kantor and Handke, while astronomy has taken over the cause of invisibility and nothingness.

Invisibility has a more banal aspect. The view of the night sky from the surface of the Earth, which had been the traditional starting point for the planetarium, has become increasingly obscured. In June 2016, a joint study by the Italian Light Pollution Science and Technology Institute and the U.S. National Oceanic and Atmospheric Administration announced that due to light pollution, 60 per cent of Europeans and 80 per cent of North Americans can no longer make out the glowing band of our galaxy, while over 30 per cent of the world's population can no longer make out the Milky Way. As illumination from streetlamps and other artificial lights and illuminations reaches into the night sky, it is reflected off moisture droplets in the atmosphere, leading to a 'sky glow'. This light pollution, which would reduce if for some reason all the electric lights in cities and on highways were switched off, is a sign of a condition that has steadily evolved since the days of Oskar von Miller.

Even as our view of the night sky is continually reduced, our space probes detect with surprising clarity ever more distant phenomena. In 1990, after many years of delays, the Hubble Space Telescope was launched; it was the first telescope to give a viewpoint from outside Earth's atmosphere. In 1992 the Cosmic Background Explorer satellite began to investigate radiation from the Big Bang; in 1997 the Cassini probe was launched to Saturn, with its lander Huygens landing on Titan in 2005; in 2012 observers using the Keck telescope in Hawaii provided the first proof of the existence of black holes; in 2015 gravitational waves from two black holes that merged 1.3 billion years ago were detected by researchers at the LIGO observatories in Louisiana and Washington State; in 2017 Voyager 2 is approaching interstellar space, and the Juno probe has psssed through Jupiter's highly unpredictable magnetic fields and commenced the first of 37 orbits. These probes can send back to Earth high-resolution images and visual scans of their field of view. They are, however, limited. Certain phenomena in the expanding universe are so remote that light from them would take too long to reach us, meaning – assuming the speed of light to be absolute – they cannot be seen from Earth. The visible cosmos does have certain limits.

How do these developments in astronomy and cosmology affect the planetarium? From dealing with a comparatively simple solar system, it now has to consider an ever-expanding universe, much of which exists beyond our visual scope. The planetarium might follow various paths, each leading in a different direction. It could become merely a kind of a museum, showing an antiquated view of astronomy, repeating its familiar shows, charming but increasingly irrelevant, like a theatre that continues to put on familiar performances for decades. It could become a kind of astronomical cinema, imitating the special effects of popular space-oriented films such as *Gravity* (2013) and *Interstellar* (2014), or even *The Tree of Life* (2011), with its entrancing analogue animations of the cosmos constructed of wax and oil. It could link to religion and spirituality, which are traditionally concerned with what lies beyond the visible. It could become more overtly technical, for with the advent of digital projectors the planetarium can adapt to continual advances in technology and the ability of computers to store vast amounts of information, allowing the ever more complex images required by contemporary astronomy to be beamed onto the hemispherical screen. Following the contemporary trend for increasingly powerful smartphones able to receive and replay visual information, one might wonder whether the planetarium might also soon become personal, producing individual versions of the digital sky.

In fact, there are already planetariums that follow each of these paths, often also intertwining one path and another, just as different types of theatre – holy, rough, immediate and so on – cross over and mix with one another. However, in almost every case the planetarium show is social, involving a mass of people enjoying together the view of the stars emerging, accompanied, even for a generation that thinks it's seen it all, by inevitable oohs and aahs of surprise as the lights go down and the stars appear.

The big change in how planetariums are able to envisage the cosmos has come through the rapid development of computer technology, and the ability to store and project vastly greater quantities of information. In our times, the digital projector supplies the equivalent of the effects Bauersfeld provided for the more confined analogue world of the 1920s.

The theme of the all-powerful computer has long been a favourite of science fiction and has featured in the planetarium since the 1960s, often being linked to that perennial favourite, the cosmic disaster story. Isaac Asimov's short story 'The Last Question' (1956) features both. Two scientists, in traditional white lab coats, discussing the end of the universe:

> 'I get it,' said Adell. 'Don't shout. When the sun is done, the other stars will be gone, too.'
>
> 'Darn right they will,' muttered Lupov. 'It all had a beginning in the original cosmic explosion, whatever that was, and it'll all have an end when all the stars run down.'

A powerful computer is asked by the two scientists whether there is any way of decreasing the entropy of the universe, which will eventually bring all life to an end. The computer replies, 'INSUFFICIENT INFORMATION FOR MEANINGFUL ANSWER'. The question is repeated at intervals over millions of years, always with the same reply. Finally, at the moment the universe dies, the computer, which now encompasses all the energy of the universe, suddenly flashes up a reply which long-vanished mankind cannot hear: '"LET THERE BE LIGHT!" And there was light.' The whole saga begins again. 'The Last Question' was a favourite theme for performances in U.S. planetariums. Versions of it, with Leonard Nimoy as the narrator, were put on in the Abrams Planetarium in Michigan, the Hayden Planetarium in New York City, in the planetariums in Edmonton, Boston, Philadelphia and many others, even to the present day. It makes sense that the story would have considerable appeal for those who run planetariums: the actual universe ends and the star show is over, but then a computer-projector, acting as a kind of mechanical prime mover, can begin the cosmic performance all over again, with its biblical declaration of the rebirth of light.

Actual improvements in projection technology evolved quickly in the 1980s, altering the nature of the planetarium show. Maryland Science Center was the first to use the All-sky system of six slide projectors, each with a customized wide-angle lens and set up to project across 60-degree triangular segments of the dome, which are then ingeniously combined to create one image over the whole 360 degrees of the dome. The images projected onto the dome were not necessarily astronomical, but could be taken from art or weather conditions, so that the dome could be used to

Planetarium laser show, Wolfsburg, 1980s.

create an illusion of being elsewhere. Images of architecture could be projected using the All-sky system, creating for instance the impression of being under the dome of St Peter's Basilica or in the starry hall of the Alhambra.

The Digistar, the first digital planetarium projector, invented by the digital graphics firm Evans & Sutherland and first installed at the planetarium in Richmond, Virginia, in 1983, projected any sequence of images produced by the computer software onto the dome through a single fish-eye lens. It was thus free from the mechanical restrictions of previous projectors – all those wonderful lamps moving at different speeds, all those different slides for what had been considered special effects, deriving from the stage tricks of the theatre, were now replaced by a simple box. Such early digital projectors had drawbacks, since they worked from a wireframe and so could only produce black and white points and lines, and the image quality was at first poor, at far lower resolution than their analogue competitors. They also lacked the presence under the dome of the great Zeiss dumbbell. But they evolved rapidly, with a wide range of manufacturers producing projectors with ever more sophisticated features and greater image resolution.

In 1993 the great Zeiss analogue dumbbell projector, which has survived in various forms since the 1920s, became a more modestly proportioned Mark VII, known as the Starball, a remarkable technical achievement that combined all its complex mechanical facilities into one roughly spherical case. Also in 1993, the new Deutsches Museum planetarium in Munich began to feature a Starball projector, with eighty single-image projectors, six video beamers and lasers on robotic arms; buttons in the seats allowed visitors to control the show. During the 1990s, shows at high-end planetariums such as those in New York, Berlin and Munich combined slide projectors, lasers, film and sound systems, all linked and controlled by computer (since the various projection devices

Zeiss Universarium digital projector.

were now too complicated for the lecturer to control manually), producing dazzling special effects: flights to distant regions of the galaxy, exploding stars, impressions of immense scale and so on. Sometimes these performances became little more than astro-discos, with spectacular laser shows accompanied by heavy metal music. For a few years the paths of stadium rock and astronomy ran in parallel in an updated version of the music of the spheres, supposed in medieval times to be played by the planets – Pink Floyd's *The Dark Side of the Moon* (1973) was launched at the London Planetarium, while the band Kraftwerk performed stadium concerts, playing their instrumental track 'Kometenmelodie' against an illuminated background of planets.

In the early years of this millennium some planetariums evolved into spectacular space-oriented entertainment centres, as much giant-screen cinemas as scientific institutes. At the time, projection equipment was very expensive, so only well-funded planetariums such as the Rose Center for Earth and Space in New York City and the Adler in Chicago were able to afford this kind of show. Today the projection equipment is cheaper, and produces high-resolution images, allowing digital projectors to become almost standard. Such projectors are now capable of ranging through time and space, of showing a view of the solar system or galaxy from any point, and are able to zoom in and out from the microscopic to the galactic. The planetarium projector can produce any desired effect, and thus to some extent can project whatever developments in astronomy are within its visual range, and can provide imaginative impressions of phenomena such as quasars and black holes. Some planetarium projectors today are linked to the Internet and can show real-time images being sent back by in-orbit telescopes and space probes, becoming in effect digital observatories for a mass audience. Rather than producing a simulation – the traditional purpose of the planetarium – real and imagined become ever more intertwined.

The rush to create digital productions has its drawbacks, however. Being able to project anything one wishes does not necessarily mean that the projected image is therefore more interesting, and a certain banality tends to set in. Many planetarium shows today are based on the projection of a pre-recorded show, often with a voiceover by a Hollywood actor, offering the universe as a space adventure as thrilling as anything seen in the cinema. Planetariums offering such shows attract large numbers of visitors as part of the contemporary science-entertainment business, which includes large-scale science museums incorporating planetariums, such as the City of Arts and Sciences in Valencia and the Cité des Sciences et de l'Industrie in Paris. However, the typical planetarium performance has also become less distinctive. The creation

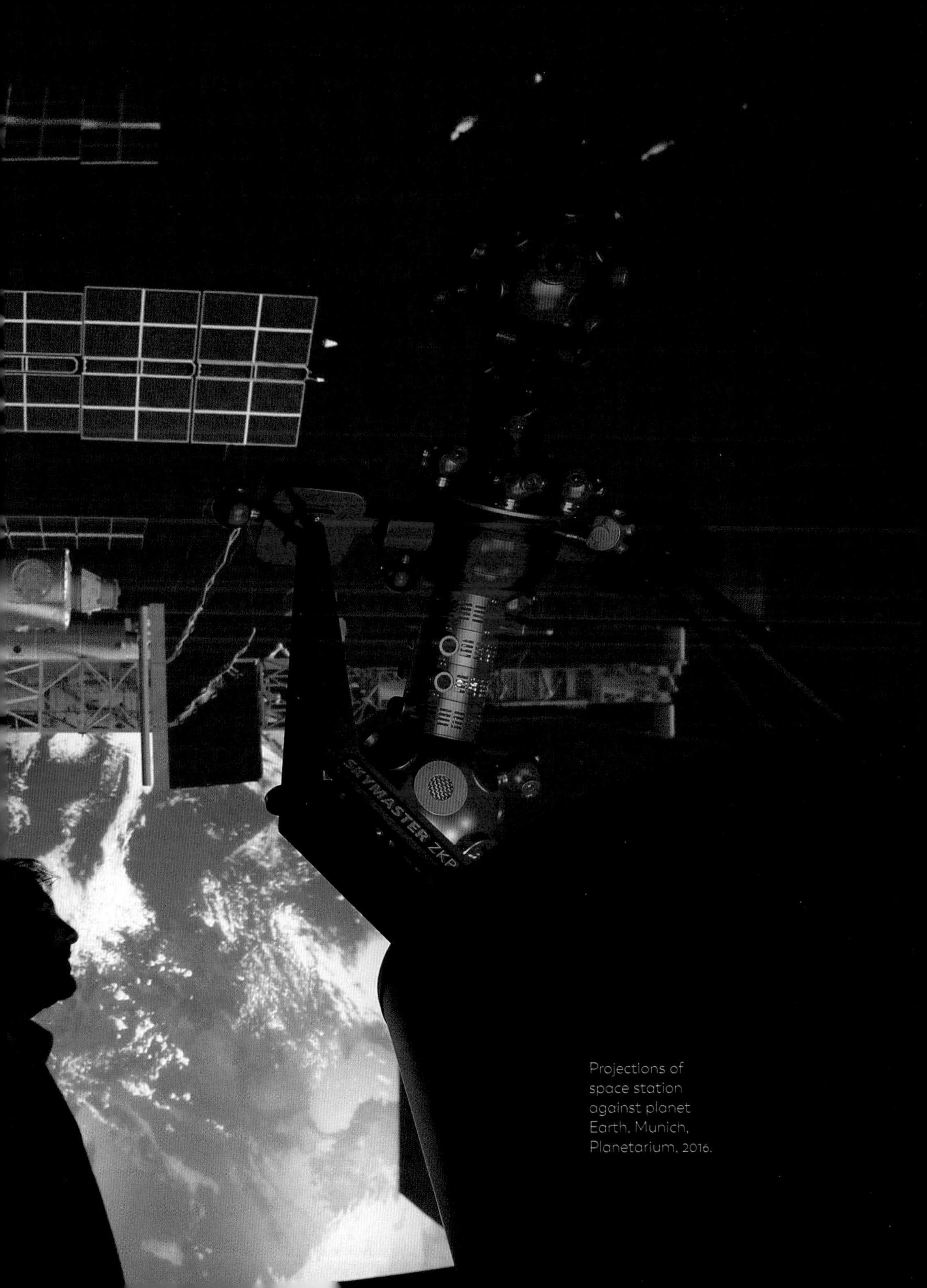

Projections of space station against planet Earth, Munich, Planetarium, 2016.

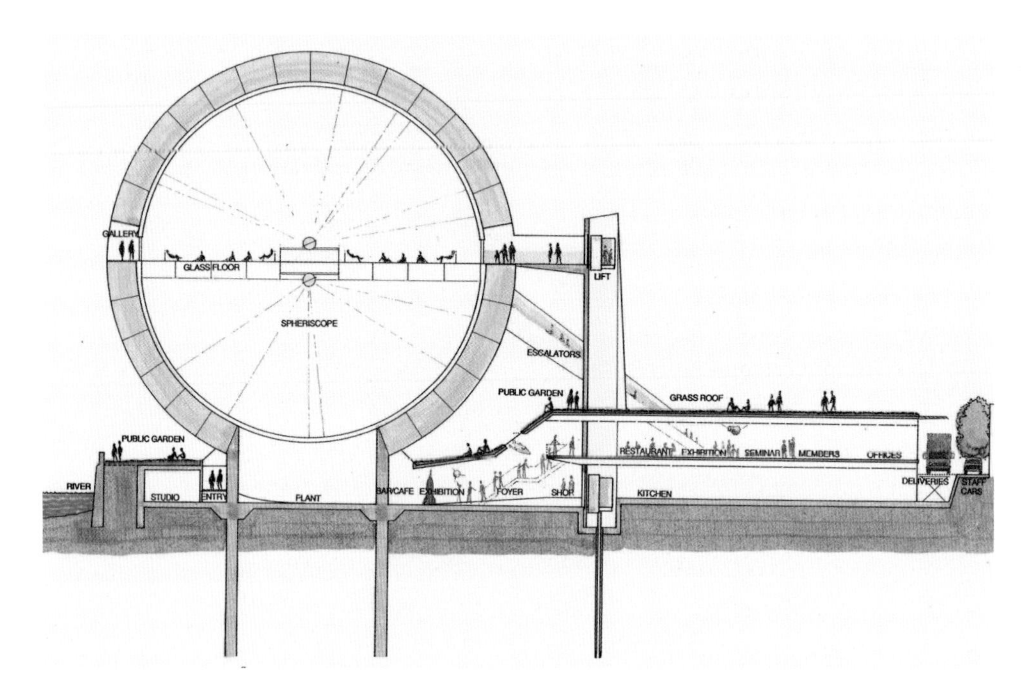

Iain Ritchie Architects Spheriscope project, section.

of the pre-recorded show presented in a standard way means the experience for the audience is already familiar: it can be seen in any planetarium in the world, whether in Shanghai, Munich or New York, and is also very similar in effect to cinema rather than theatre, which is live, always varying, and responsive to the atmosphere provided by a changing audience. Many planetariums have moved away from the traditional circular seating pattern around a centralized projector to banked rows of seats, with the audience all facing in one direction, and what is called a tilt-dome screen, whereby the hemispherical dome is tilted at 167 degrees to make it easier for the audience to see from their seats. Where in all this is the sense of theatre? Some contemporary planetariums are often nearer to the deadly theatre, providing thrills but also being predictable and impersonal.

But the heavens tolerate no terrestrial mediocrity. The spirit of boldness and experimentation of the original planetariums still survives. A full-dome, as the interior of the contemporary planetarium is known, is in itself uninteresting, little more than a rather average cinema. However, what if, instead of keeping to the traditional dome, the actual space within the planetarium was altered? After all, the basic notion of the dome as an artificial sky dates from the days of King Khosrow. Étienne-Louis Boullée's great eighteenth-century globe had suggested a completely spherical interior, but remained unconstructed. There have been various proposals for 360-degree planetariums, with computers capable of projecting high-resolution visual information onto their interiors. In 1985 the London-based Ian Ritchie Architects proposed a Spheriscope to be located on the riverside at Greenwich, not far from the prime meridian. The Spheriscope would be a 30-metre-diameter glass sphere, containing in turn a steel sphere with a non-reflective glass floor inserted across its diameter. The external sphere would be held in place by peripheral external columns. Three hundred spectators would be able to lie or sit on the glass floor, and two projectors – one for the upper hemisphere and one for the lower – would beam onto all 360 degrees of the sphere's interior, giving an illusion of total space.

One of the architects, Simon Conolly, describes the sensation:

> Spectators would be given soft overshoes on entry. The young/fit/able-bodied could sit anywhere they like on the glass floor, those of a more cautious disposition could sit on seats around the periphery. Where to sit in a planetarium is always an interesting question, and the traditional advice is to see which way the presenter's control desk is facing. But with images projected onto both upper and lower hemispheres, with just a narrow blank band at the equator to avoid dazzling spectators,

Hayden Planetarium, Rose Center, New York, 2000.

the sense of immersion would be quite dramatic. We consulted neurological/perception experts at UCL, who confirmed the experience of space would be very powerful.

The architects were advised that they would have to allow for the consequences of motion sickness, brought on by the visitors' experience of being seemingly weightless within the continuously moving projection on the internal skin of the sphere.

At the time of the Spheriscope's design, still in the very early years of digital projection technology, there were no computers with adequate processing power to create suitable pixel resolution for such a large internal space. But computer capacity has developed so much in the decades since that this would now be less of a problem. This enterprising project, which is far bolder than any contemporary digital special effects planetarium, sadly did not receive funding.

The most interesting aspects of the contemporary planetarium are what happens on the interior, with the effects of the show, rather than with the external form. An immense space, of uncertain dimensions, is created through the projection of light. This dematerialization of space was one of the ambitions of early twentieth-century architecture, with its proposals for glass buildings with minimal interior divisions. The planetarium show goes much further, suggesting that material boundaries might be dissolved altogether – that those in the audience are located both in their own everyday space and in the apparently endless space of the cosmos. The boldness and ambition of this proposal has rarely been matched by the exterior architecture of the planetarium, which tends to be an enclosure adapted to one of a variety of architectural styles. To some extent, whatever occurs within the hemispherical interior dome can be fitted into most architectural forms – a sphere, a temple, a modernist, postmodernist, neomodernist or whatever-is-currently-in-fashion volume. What kind of planetarium architecture might be inspired by the non-visual, immaterial astronomy of our times is a question with no clear answer.

While no planetarium has been bold enough to follow the lead of the Spheriscope to produce a spherical internal projection space, there are nevertheless numerous planetariums that work with the sphere shape, which as an external architectural form is naturally reminiscent of planets and other celestial bodies.

The Rose Center for Earth and Space in New York is the most spectacular of these planetarium spheres and contains, among various other facilities, the new Hayden Planetarium. Although it opened in 2000, it remains a state-of-the-art centre for contemporary astronomy, illustrating not only the advantages of but the questions surrounding the latest generation of planetariums. The old Hayden Planetarium, dating back to 1933, had been updated at various times with new projectors and other equipment, but by the mid-1990s the building itself was considered too outmoded to convey contemporary astronomical notions. After much discussion, since the Hayden was considered a New York landmark, it was demolished in 1997 and replaced by a new building, designed by Polshek Partnership Architects, backed up by various teams of media consultants and exhibition designers. Whereas most planetariums are enclosed boxes, since they require no daylight, and usually have no windows, the Rose Center offers a translucent box containing an opaque sphere. From the exterior, the design is a demonstration of the idea of the planetarium as a planet, as the large, grey sphere, 27 metres in diameter, appears to float within the six-storey-high glass cube, and is brightly illuminated at night, like a new celestial body hovering over the streets of New York. It is difficult to design a sphere that can be suspended in space or sit comfortably on the ground, and when one enters the building one realizes that the Rose Center sphere is in fact supported by substantial columns, a structural solution rather at odds with the idea of the free-floating planet. The forces of gravity that determine the attraction between planets thus lead to pragmatic engineering to keep the vast bulk of the Earth and the comparatively small volume of the Rose Center sphere apart.

The upper half of this sphere is filled by the Star Theater, with a tilt-dome, a one-off Zeiss digital projector – the Mark IX Hayden – and a highly sophisticated projection system producing a hyperrealistic animated view of the cosmos. The pre-recorded thirty-minute shows are usually narrated by Hollywood celebrities, for example 'Passport to the Universe', narrated by Tom Hanks; 'Journey to the Stars' with Whoopi Goldberg and 'Dark Universe', narrated by the planetarium's director, Neil deGrasse

Tyson, himself. Effectively, the Star Theater is a kind of cinema, offering ever more dramatic portrayals of the cosmos.

'I'm baffled all the time,' states Tyson, with unaccustomed uncertainty for an astronomer, 'We don't know what's driving 96% of the universe. Everybody you know and love and heard of and think about and see in the night sky through a telescope: four per cent of the universe.' In fact, the Rose Center is filled with demonstrations of the physical aspects of the universe. Beneath the Star Theater, occupying the lower half of the great sphere, is the Big Bang Theater, which features a programme explaining the origins of the universe. The Heilbrunn Cosmic Pathway forms a walkway that winds around the sphere, so that 1 foot represents 36,111,111 years, the visitor walking from the Big Bang up to the extinction of the dinosaurs. Large models of the eight planets – Pluto is controversially excluded as a pseudo-planet – are suspended beside the large sphere.

Among the most interesting features of the Rose Center is the Digital Universe Atlas, where 'every satellite, Moon, planet, star and galaxy is represented to scale and in its correct, measured location according to the best scientific research to date.' All the known celestial bodies of the universe have gradually been catalogued in a three-dimensional index. Only a computer of vast capability, similar to the one in Asimov's story, could begin such a project, including the endless task of keeping the atlas up to date. At present, the Universe Atlas is displayed as a fly-through projection, in which the spectator appears to move through the stars, but one could imagine it developing into some kind of 3D matrix, where the viewer is placed within the system and can roam within it. The ambition of this Universe Atlas could then put into question how the display of visual information could form a special relationship with the spectator. Where would the actual spectator be located in relation to the digital index? How would one move through it? How

Alexandria Planetarium, 2009.

would the non-visual be visualized? All such questions suggest the evolution of a new kind of star theatre, yet to be realized.

The new Hayden Planetarium is theatrical in the sense that it presents the great sphere of the planetarium as a brightly illuminated object, to be put on display to the surrounding city, with the visiting human beings reduced to a miniature scale. The planetarium as a dome is a traditional solution and matches the internal hemispherical projection screen – it has a clear structural rationale. But the sphere is less clearly rational as an architectural form, for spheres tend to roll around; the structure needs to sit on a single point; and the lower part of the interior space is hard to use. In spite of these drawbacks, the sphere is an attractive form, simple and strong, relating back to the old notion of the solar system as a set of celestial spheres. Since the late 1980s the sphere has almost become one of the standard formal solutions to the question of what a planetarium should look like: a planetarium looks like a miniature planet. Planet Earth is gradually becoming invaded by a smaller breed of artificial planets, with which are displayed the wonders of the cosmos – projected planets-within-planets, constructed in three dimensions.

The Paris planetarium in the Cité des Sciences et de l'Industrie, opened in 1986, was one of the earliest of these grand spheres – a 22-metre, free-standing metallic globe, clearly again resembling a miniature artificial planet, sitting between the museum and the Parc de la Villette. It has a certain formal authority and is like an oversized toy; one wonders just what might be inside. The Bristol planetarium (opened 2000), which is also attached to a museum, is very similar, if smaller in scale. The very interesting planetarium at the Bibliotheca Alexandrina in Egypt, which opened in 2002 and was designed by the Norwegian-American architecture practice Snøhetta, is a moody, black sphere, visually divided into sections by vertical lines and partly sunken into a pit in a concrete plaza. It resembles a dark planet emerging out of the time of the goddess Nut.

Other spheres, or sometimes three-quarter spheres, appear in the large number of planetariums in China, including in Beijing, Fuzhou and Xiamen; and in Japan, including Kanagawa and Osaka. The Inner Mongolian Planetarium in Hohhot is a playful, bright orange sphere with horizontal stripes (a decoration as yet unknown in the solar system), and is placed beside a large undulating roof, as though about to be carried away by a cosmic storm. But the largest and most spectacular of these spheres is in Nagoya, Japan, a vast, silvery form that appears to float freely above the ground but which is in fact supported by the two adjoining buildings. The Nagoya sphere, built in 2010 with a diameter of 35 metres and accommodating 350 spectators, is the largest planetarium anywhere – a paternal Jupiter to all other globes. Sadly, all these spheres are a matter of exterior form only, and none follows the logic of the Spheriscope or of Boullée's cenotaph in using the interior as a spectacular spherical space. We still await the truly spherical planetarium.

Two smaller spheres are worthy of mention. The planetarium in Valletta, Malta, which opened in 2016, is a sphere curiously set on the ruins of a historic building, as though having recently arrived at speed from space. The Valletta planetarium is a cheerful and rather subversive harbour-side object, combining scientific education with seaside entertainment. Rather more subtle is the Infoversum (2014) in Groningen in the Netherlands, a small-scale planetarium-cinema, initiated by the astronomer Edwin Valentijn and designed by Jack van der Palen of the architects Archiview. It consists of a white sphere sitting on a curved base of Corten steel and, as suits a building based on a planet, is only in a temporary location: it could be moved elsewhere. The building is light and delightful, a Mercury rather than a Saturn or Jupiter, and keeps its options open as to just what its architecture symbolizes – it could be a celestial object, a hat, a ship or an oversized egg cup. Within, there is a rather conventional cinema layout. Unlike many spherical planetariums, which work too hard at being imposing, it does not take itself too seriously and keeps the audience slightly baffled as to its intentions.

'Further on, nothing'? For the most part, these new planetariums, whatever their exterior form, present a show illustrating a fairly standard view of the cosmos, based on current Western science. Are all planetariums still scientific? Yes, but science is easily adapted to other purposes. As noted earlier in the observation by Frank Close on the nature of nothingness, contemporary scientific theories emerge from very ancient considerations of the nature of the universe, including early Indian texts.

'Then even nothingness was not, nor existence', runs the highly mystical Vedic Nasadaya Sudka, or Hymn of Creation: 'There was no air then, nor the heavens beyond it. What covered it? Where

Nagoya Planetarium, 2011.

Cafe &
Restaurant

was it? In whose keeping? Was there then cosmic water, in depths unfathomed?' India has also continued to offer the most individual take on the architecture of the planetarium, without necessarily involving the convolutions of the Nasadaya Sudka, but acting as a balance to the tendency in both the West and East to reduce the planetarium to a technological artefact. Each Indian planetarium appears to have started from a different point of view, as though with the assumption that no fixed form preceded it and going on instead to invent one. The Pushpa Gujral Planetarium in Kapurthala in north India, opened in 2005, is a large Pop art Earth sphere boldly marked out with the continents. The Indira Gandhi Planetarium in Lucknow (2003) takes the planet comparison to extremes. Located among a group of mundane housing blocks, it borrows the form of the planet Saturn: an orange globe complete with a set of orbital rings, the sphere appearing to be still spinning, as though about to return to its proper position in the solar system. The Indira Gandhi Planetarium is ingenious and amusing.

Groningen Infoversum Planetarium, 2014.

Why would Saturn, traditionally the gloomy planet, have landed in Lucknow, an ancient town on the edge of the Himalayas now grown to a modern city – is it some portent of an unexpected event that will occur in the near future?

Almost all Indian planetariums operate with a standard view of astronomy, little different from that presented in the West. There are interesting exceptions, however. The planetarium in the Sri Sathya Sai Institute of Higher Learning, in the ashram at Puttaparthi in southeast India, is part of a diverse set of buildings dedicated to the guru Sathya Sai Baba. Sai Baba claimed to be the incarnation of an earlier guru, Sai Baba of Shirdi, and had a considerable reputation as both a miracle worker, able to materialize material objects, and as an ingenious conjurer – depending on one's view of performers of miracles. Sai Baba was certainly a philanthropist on a grand scale and an enthusiast for diversity in architecture. The ashram, largely completed after his death in 1994, features, among many other buildings, a large hospital resembling John Nash's Orientalist fantasy, the Royal Pavilion in Brighton; the Chinese-Moorish-Gothic Chaitanya Jyoti Museum; and the extraordinary Sanathana Samskruthi spiritual museum, a large, white volume vaguely reminiscent of Shaker architecture.

The Sri Sathya Sai Space Theatre, constructed with funds donated by a U.S. follower for Sai Baba's sixtieth birthday celebrations, and inaugurated by Sai Baba himself in 1985, celebrates, with a Pop art sense of styling, astronomy in a spiritual environment, and takes the form of a funky, faceted, multicoloured dome, the facets painted boldly in green, yellow, white and mauve. Once again we have an unlikely mix of architectural influences – it is Buckminster Fuller meets Jeff Koons, meets traditional South Indian architecture. One might recall that phrase of Alice Munro, that the planetarium is already a 'slightly phoney temple'; here the temple aspect, which seems to lie somewhere in the shadows in contemporary planetariums, now emerges into the light. In the interior of the Sri Sathya Sai Space Theatre, the constellations with their traditional symbolism, navigational stars, globular clusters, the Milky Way, Magellanic Clouds and nebulae are projected via the most advanced digital projectors, accompanied by the narration of myths from ancient Sanskrit texts and films celebrating the miracles reputedly performed by Sai Baba. The planetarium thus ingeniously mixes together a sense of the absurd with both contemporary digital projection technology and the requirements of a spiritual approach to the cosmos. Compared to the seriousness of many contemporary planetariums, there is something very enjoyable in this approach.

Even the Sri Sathya Sai Space Theatre is put in the shade by the extraordinary Temple of the Vedic Planetarium, currently under construction in Mayapur in West Bengal, India, under the supervision of the New York-based spiritual association ISKON, the International Society for Krishna Consciousness (more commonly known as the Hare Krishna movement). The planetarium temple is a construction of ambitious size located in the flat fields of Bengal. Its exterior style boldly blends together elements of the United States Capitol building, Art Deco, Indian architecture, Fabergé style decoration, Disneyland and any number of other stylistic inspirations, to produce extended facades with bright blue panels, towers and a formal entrance, all surmounted by a 50-metre-diameter dome, claimed to be upon completion the largest in the world. The intention is for the building to act both as a temple for pilgrims and as a planetarium dedicated to demonstrating the principles of Vedic cosmology. These principles, demonstrated by a large-scale, highly detailed physical model of the universe as described in the Fifth Canto of the *Srimad Bhagavatam*, are of considerable interest, offering an alternative to the materialist approach of Western science. The Vedic Planetarium website declares:

Sri Sathya Sai Space Theatre, Puttaparthi, 1985.

> Cosmology is defined as the study of the origin, purpose, structure and functioning of the universe. Vedic cosmology gives a large amount of information about not only the structure of the phenomenal universe as we see it, but also a clear idea of the source of the manifested universe, its purpose, and the subtle laws that govern its operation.

How the universe manifests itself and from where it originates are questions that should be asked in any planetarium. Within the Vedic Planetarium dome will be a large chandelier, a physical demonstration of Vedic cosmology. Any attempt to describe this immensely complicated chandelier would be reductive. However, it rises through numerous levels and includes numerous figures from Vedic myths, the seven subterranean planets, a central island with a mountain, golden mountain cities and a mandala with additional islands, as well as moving elements such as the Moon and Sun (which appears to orbit the Earth) on chariots, the zodiac of stars, the planet of the seven sages and the 24 additional planets. Judging from online videos, the chandelier looks like it will be a

Temple of the Vedic Planetarium, Mayapur, 2017.

wonderful construction, much more convincing than the exterior architecture of the building. But to what extent these elements should be represented visually, and to what extent they are in fact metaphorical, are questions whose answers would no doubt be long and complicated. One might also ask whether this is simply a model or if there are other notions involved – does it seek in some way to influence the world, to be active rather than passive? And where does the observer stand in relation to the model – on its exterior, looking in from the building's layers of balconies? Or, is our position to such a recreation of a spiritual cosmos really very different, we cannot merely remain spectators but inevitably become part of the show.

In another part of the Vedic temple there will be a more conventional planetarium, with a digital projector beaming onto a hemispherical screen images similar to those found in Western planetariums, but of course adapted to Vedic principles. It remains unclear just how the differences between Vedic and current 'scientific' cosmology will be dealt with in this planetarium – for instance, how many planets are presented, and whether the Sun travels around the Earth, or vice versa. Much of this depends on just how these models are to be understood – are they intended to show a physical or mystical world?

The Temple of the Vedic Planetarium recalls the planetarium built in 1930s San Jose, with its mix of Bauersfeld projection technique and Rosicrucian cosmology, but now evolved into digital technology and Vedic cosmology. The Vedic Planetarium in Mayapur acts as a balance to the new Hayden Planetarium in New York; each is concerned with drawing masses of visitors, employing a vast architectural scale to impress, using a mixture of representation techniques ranging from physical models to the most advanced digital technology, proposing an understanding of the universe as spectacle, and combining elements of what is comprehensible with large doses of mystery and wonder. Despite their wondrous complexity, each omits the modest and emotive quality of the direct relationship of the night sky to a human being on the surface of the Earth.

There are other alternatives to the standard planetarium. A scale model of the night sky can itself exist outside, under the actual night sky. Out in the Nevada Desert, a series of cars moves slowly at night across the sands, with their lights illuminated, each following one of a series of concentric paths, spaced out to match the scale of the distances between the actual planets. The medieval system of interlocking spheres is reduced to two dimensions for the automobile age. This is the experimental planetarium called *To Scale: The Solar System*, which was constructed in 2015 by a small group of LA film-makers including Wylie Overstreet and Alex Gorosh. As the name suggests, the project is concerned with conveying the vast scale of the distances between the planets, which remains undemonstrated in exhibits like Miller's Copernican planetarium or the Rose Center's Six Scales, because while the planets are at the correct scale in relationship to each other, the enormous distances between them are necessarily very much reduced. 'If you put the orbits to scale on a piece of paper,' states Overstreet on the online video illustrating the project, 'the planets become microscopic, and you won't be able to see them. There is literally not an image that adequately shows you what it actually looks like from out there. The only way to see a scale model of the solar system is to build one.'

The astronaut James Irwin described planet Earth as appearing 'the size of a marble, the most beautiful marble you could imagine'. Using the marble-sized Earth as a starting point, the Sun becomes 2.5 metres in diameter, and the other planets assume their appropriate size. To construct a model of the solar system at this scale requires a space over 11 kilometres in diameter; Overstreet and Gorosh found it in a dried-out lakebed in the Black Rock Desert, Nevada. The group worked out the appropriate size of the planets and their distance from one another. Mercury is 64 metres from the Sun, Venus 120 metres, and Earth 176 metres, while Uranus is 3.4 kilometres away and Neptune 5.6 kilometres. The orbit of Pluto, which has been much maligned and recently reduced to the official status of a dwarf planet, would unfortunately not fit into the space in the desert.

The experiment lasted for 36 hours. The desert is empty apart from the cars. There is no audience on site, only online. Somewhere up in the mountains above the circling cars, a cameraman awaits to record their movements with time-lapse recording – the camera position is pragmatic, up on the mountain, an easy position from which to view and film the circling cars but outside the scaling system. On the resulting video the lights of the cars, following their allotted planetary paths – which look, however, more circular than elliptical – shine out, creating the effect of rings of illumination, matching the scale of the distances between the actual planets. Unlike in a conventional planetarium, it is these automobile planets that are now providing the illumination, rather than a central

To Scale: the Solar System planetarium installation, Nevada Desert, 2016.

projector. The resulting video, which has attracted 1.5 million hits on the Internet – a number with which any planetarium director would be delighted – gives a remarkable demonstration of the scale of the solar system, effectively flattened to two dimensions.

One can move rapidly from the scale of the desert to that of a hand-held device. On 14 July 2015, after a journey lasting nine years, NASA's New Horizons spacecraft flew past Pluto and its moons, scanning the dwarf planet with its various spectrometers, radiometers, cameras and other instruments, and sending the data back to Earth. A year later, the *New York Times* offered a free app for mobile phones, on which the data from the spacecraft could be observed. 'Watch New Horizons glide through space at a million miles a day,' runs the portentous commentary, set to spooky space music; 'Fly over Pluto's rugged surface and smooth, heart-shaped plains. Stand on icy mountains as the moon Charon looms on the horizon. Touch down in a frost-rimmed crater, billions of years old.' The space probe's images are more intriguing: one can see the planet, and then zoom in down to an image representing a space less than 100 metres wide, to move across the craters and rocks of this extraordinary icy landscape. In fact, the app is for the most part a digital animation, supplying an ingenious but largely invented virtual version of the original data, which appears to be sent direct from the spacecraft but which has actually been recreated by media experts. Perhaps soon such fakery will not be required, however, and we will indeed be able to receive such data directly (with the obvious delay due to the time it takes for the signal to arrive). This will not really offer any replacement for the planetarium, which traditionally provides a model, a simulation, while an observatory offers a direct view. But, since the planetarium exists in the first place because a direct view of the sky from the surface of Earth is for many people no longer possible, one could imagine that some app in the near future could relate the direct visual data from a probe to a broader simulation, and thus become a miniature, hand-held planetarium.

'Let there be light!' might be the cry not of a highly sophisticated computer, but of the simplest technology. Take U.S. Highway 8, which leads east out of Monico in Oneida County, Wisconsin, with its population of 364 at the last census and its astronomical topography of lakes named after Mars, Venus and Neptune. After a couple of kilometres, you take a turning north down Mud Creek Road, and pull over at the plot numbered 2392. You get out, stand on the soil, breathe in the fresh Wisconsin air. Is there anything much to expect out here? A few trees and meadows, some sheds . . . the roads are empty. You recall the advice of Henry David Thoreau, that lone woodsman of Massachusetts, a believer in the direct appreciation of the natural world as opposed to any understanding through technology: 'Heaven is under our feet as well as over our heads.' This could now perhaps be modified to: Heaven is under our feet as well as on a hand-held device. On the edge of a clearing stands a white timber-frame shed, with a pitched roof, painted white: yet another woodland barn, or a primitive temple in the woods. The words Kovac Planetarium are written out over the entrance. You open the door of the shed, and allow your eyes to adapt to the low light.

Within is a large timber sphere, hand-built by Frank Kovac, a worker in the local paper mill. Kovac is usually there to greet visitors, as this is his personal planetarium, his own version of the night sky. As a child, living in Chicago and visiting the Atwood Planetarium, he had always wanted to be an astronomer, but could not do the maths, so he settled down to work in the mill. Later, as a teenager in Wisconsin, Kovac had a defining experience, of seeing – or rather, not seeing – the stars. In his own words:

> The year was 1996, and it turned out to be a beautiful clear October day. Hopes were high as a group of Boy Scouts were eager to spend the evening under a star-studded sky at Mud Creek Observatory. Just after sunset excitement soon turned to disappointment as cloud cover rolled in obscuring the universe once again. This was the night my dream was born. I would take matters into my own hands and build a planetarium.

Kovac wanted to create a space that would demonstrate the wonder of the night sky, which would not be obscured by unexpected weather conditions and which could be created at a scale he could achieve on his own. He could have bought a small-scale Spitz planetarium projector, but this would have been too impersonal, too ordinary. So he built his own version of the night sky. Realizing that he did not need a projector at all, he constructed out of timber ribs and plywood sheets a three-quarter sphere 7 metres in diameter, driven by an electric motor so that it rotates smoothly at the 22.5-degree angle of the ecliptic, the line marking the apparent path of the sun across the sky over the course of the year. There is space inside the sphere for a dozen spectators.

You sit down on one of the wooden benches. The interior of the planetarium is painted black, and on the surface Kovac has painted by hand 5,200 stars of the northern hemisphere. They are in their correct positions for this location – if in doubt, Kovac slipped outside to check on the actual sky – and allowing for differences in brightness to create the impression of depth. These stars painted in luminous paint are charged up with an electric lamp, and then glow softly for eight hours when the lights go down. The sphere rotates gently and silently, fading into immateriality as the luminous stars emerge. Moveable planets are positioned by hand, gradually moving along their allotted paths. Kovac gives a running talk about the constellations and planets – he is another in that

Frank Kovac planetarium, 2008.

Frank Kovac planetarium, 2008.

long line of planetarium lecturers who offers a personal view of the universe. There's no dark matter here, no parallel universes or space-time conundrums. In no way does Kovac's planetarium compete with today's highly sophisticated digital planetariums; it is in a completely different category, echoing a time before even Bauersfeld's invention. Holy – Rough – Immediate – Spherical – Mechanical? A touch of each. The Kovac works has the quality of theatre, which is always live and unpredictable as opposed to pre-recorded and fixed, and involving a human being in direct relationship with the audience, in a space that is both small and vast. The best planetariums are in some way personal; they belong to a person or small group. They are not mere machines, however ingenious.

At the end of the show the lights are raised and the stars fade away. You move out through the shed to the small clearing in the woods. You gaze up at the sky, which is conducting its own slow and regular performance. Is the DIY planetarium opposed to the most advanced contemporary technology? You might also consult

Projection of star constellations onto stellar background, Stuttgart Planetarium, 2016.

on your smartphone the latest images sent back by space probes, receiving with appropriate delay an actual view of planets you can never expect to reach. A slowly revolving hand-painted planetarium, a view of the actual night sky, and a digital image of a distant celestial body. The heavens are as full of light as ever. We live easily enough with the contradictions inherent in the different theatres of the stars.

TIMELINE OF SELECTED PLANETARIUMS

For a more inclusive list, see www.aplf-planetariums.info/en/index.php

c. 530	Persia (Palace of Khosrow)
c. 740	Qasr Amra, Jordan
c. 1325	Yazd, Iran (Dome of Jameh Mosque)
c. 1500	Granada, Spain (Alhambra ceiling of Hall of the Abencerrajes)
1588	Florence, Italy (Santucci sphere)
1654–64	Gottorf, Germany (Gottorf sphere)
1661	Jena, Germany (Weigel sphere)
1730–38	Delhi, India (Jantar Mantar Observatory)
1772	Cambridge, England (Roger Long sphere)
1774–81	Franeker, Netherlands (Eisinga Orrery)
1784	France (Étienne-Louis Boullée, cenotaph for Isaac Newton, unbuilt)
1794	France (Jean-Jacques Lequeu, Temple de la Terre, unbuilt)
1816	Berlin, Germany (Karl Friedrich Schinkel, Queen of Night stage set)
1838–43	Strasbourg, France (astronomical clock, Schwilgué version)
1851	London, England (James Wyld globe)
1900	Paris, France (Élisée Reclus globe, unbuilt)
1913	Chicago, USA (Atwood Globe)
1923	Munich, Germany (Deutsches Museum)
1923	Jena, Germany (Zeiss rooftop, Bauersfeld Sternentheater)
1925	Sugarloaf Mountain, Maryland, USA (Frank Lloyd Wright proposal, unbuilt)
1926	Wuppertal/Barmen, Leipzig, Düsseldorf, Jena, Dresden and Berlin, Germany
1927	Mannheim and Nuremberg, Germany; Vienna, Austria
1928	Hannover and Stuttgart, Germany; Rome, Italy

1929	Moscow, USSR
1930	Hamburg, Germany; Stockholm, Sweden; Chicago, Illinois (Adler); Milan, Italy; Vienna, Austria
1933	Philadelphia, Pennsylvania (Fels)
1934	The Hague, Netherlands; Springfield, Massachusetts (Korosz projector)
1935	Los Angeles, California (Griffith); Brussels, Belgium; New York City (Hayden)
1936	San Jose, California (Rosicrucian Park)
1937	Osaka, Japan; Paris, France
1938	Tokyo, Japan
1939	Pittsburgh, Pennsylvania
1949	First Spitz projectors
1954	Stalingrad and Penza, USSR; Pune, India
1955	Katowice, Poland
1957	Peking, China; São Paulo, Brazil
1958	London, England
1960	Prague, Czechoslovakia; Lima, Peru (Morro Solar)
1963	Reno, Nevada (Fleischmann Atmospherium); St Louis, Missouri (Science Center)
1964	Belgrade, Serbia
1965	Colombo, Sri Lanka
1966	First Stern portable projector, with inflatable dome; Buenos Aires, Argentina (Galileo Galilei)
1967	Calgary, Canada
1968	Toronto, Canada
1974	Cottbus, Germany (Raumflugplanetarium Juri Gagarin)
1977	New Delhi, India (Nehru Planetarium); Budapest, Hungary
1981	Tripoli, Libya
1983	Richmond, Virginia (first Evans & Sutherland Digistar 1); Wolfsburg, Germany; Wanangal, India
1984	Edmonton, Canada
1985	Puttaparthi, India (Sri Sathya Sai Space Theatre)
1986	Paris, France (Cité de Science et de l'Industrie)
1987	Prenzlauerberg (Zeiss-Grossplanetarium), and Berlin, Germany
1989	Bangalore, India (Jawaharlal Nehru Planetarium)
1998	Valencia, Spain (L'Hemisfèric, City of Arts and Sciences)
2000	New York City (Hayden Planetarium)
2003	Mumbai (Nehru) and Lucknow (Indira Gandhi), India
2007	Greenwich (Peter Harrison), London
2010	Monico, Wisconsin (Kovac)

2012	Dome of Mina Tehran, Iran; Hohhot, Mongolia
2015	Black Rock, Nevada (*To Scale: The Solar System*)
2018	Mayapur, India (Temple of the Vedic Planetarium, under construction)

PRINCIPAL PLANETARIUMS OF ARCHITECTURAL INTEREST

The nature of which buildings are of architectural interest is naturally an open one; this list has been compiled by the author. For a more inclusive list of all planetariums, see www.aplf-planetariums.info/en/index.php

AFRICA

Egypt

Alexandria	Planetarium Science Center, Bibliotheca Alexandrina (2002)

Ghana

Accra	Ghana Planetarium (2009)

Libya

Tripoli	Tripoli Planetarium (1981)

Tunisia

Tunis	Cité des Sciences (1996)

ASIA

China

Guangzhou	Space Flight Spectacle (2005?)
Hohhot	Inner Mongolia Science and Technology Museum (2012)
Shanghai	Shanghai Planetarium (under construction, due to open 2018)

India

Bhubaneswar	Pathani Samanta Planetarium (1990)
Delhi	Nehru Planetarium (1984)
Jaipur	B. M. Birla Planetarium (1989)
Kolkata	B. M. Birla Planetarium (1962)
Lucknow	Indira Gandhi Planetarium (2003)
Mayapur	Temple of the Vedic Planetarium, under construction
Mumbai	Nehru Planetarium (1977)
Porbandar	Shri Jawaharlal Nehru Patel Planetarium (1976)
Puttaparthi	Sri Sathya Sai Space Theatre (1985)
Vadodara	Sardar Vallabhbhai Patel Planetarium (1976)

Japan

Nagoya	Nagoya City Science Museum (1962, rebuilt 2010)

Malaysia

Kuala Lumpur	Planetarium Negara (2000)

Sri Lanka

Colombo	Sri Lanka Planetarium (1965)

EUROPE

Belgium

Brussels	Planétarium de l'Observatoire Royale de Belgique (1935)

Czech Republic

Prague	Planetárium Praha (1960)

Denmark

Copenhagen	Tycho Brahe Planetarium (1989)

France

Paris	Palais de la Découverte (1952)
Paris	Cité des Sciences et de l'Industrie (1986)

Germany

Berlin	Zeiss Grossplanetarium (1987)
Cottbus	Raumflugplanetarium Cottbus (1974)

Garching	ESO Supernova Planetarium (under construction, due to open 2018)
Hamburg	Planetarium Hamburg (1930)
Jena	Zeiss Planetarium (1926)
Stuttgart	Carl-Zeiss-Planetarium (1977)
Wolfsburg	Planetarium Wolfsburg (1983)

Greece

Athens	Eugenides Foundation Planetarium (1966)

Hungary

Budapest	TIT Budapesti Planetárium (1977)

Malta

Valletta	Esplora Interactive Science Centre (2016)

Netherlands

Groningen	Infoversum (2014)

Poland

Chorzow	Planetarium Słaskie (Silesian Planetarium) (1955)
Warsaw	Centrum Nauki Kopernik (Copernicus Science Centre) (2011)

Russia

Kaluga	Konstantin E. Tsiolkovsky State Museum of the History of Cosmonautics (1967)
Moscow	Moscow Planetarium (1929)
Penza	Planetarium Penzenskogo (1954)
Volgograd	Volgograd (formerly Stalingrad) Planetarium (1954)

United Kingdom

Glasgow	Glasgow Science Centre Planetarium (2001)
London	London Planetarium (1959, now disused)
London	Peter Harrison Planetarium, Greenwich Observatory (2007)

Spain

San Sebastian	Eureka! Zientzia Museoa Planetarium (2001)
Valencia	L’Hemisfèric, City of Arts and Sciences (1998)

Turkey

Konya	Konya Bilim Merkezi (2014)

NORTH AMERICA

Canada

Edmonton	Margaret Zeidler Star Theatre (1984)
Quebec	Rio Tinto Alcan Planetarium (1966)
Toronto	McLaughlin Planetarium (1968, closed 1995)
Vancouver	Star Theatre, H. R. MacMillan Space Centre (1968)

USA

Chicago	Adler Planetarium (1930, additions 1997 and 2010)
New York	Hayden Planetarium, Rose Center for Earth and Space (1935, reopened 2000)
Los Angeles	Samuel Oschin Planetarium, Griffith Observatory (1935)
San Jose	Rosicrucian Park planetarium (1936)
St Louis	St Louis Science Center and Planetarium (1963)

Mexico

Ciudad Victoria	Planetario Dr Ramiro Iglesias Leal (1992)

SOUTH AMERICA

Argentina

Buenos Aires	Planetario Galileo Galilei (1967)

Brazil

São Paulo	Planetário de Ibirapuera Prof. Aristóteles Orsini (1957)

Colombia

Bogotá	Planetario de Bogotá (1969)

Peru

Lima	Planetario José Castro Mendívil, Observatorio Morro Solar (1960)

Uruguay

Montevideo	Planetario Municipal Agr. Germán Barbato (1955)

FURTHER READING

1 HOLY, ROUGH, IMMEDIATE

Atwood, Wallace W., *The Atwood Sphere* (Chicago, IL, 1913)

Brook, Peter, *The Empty Space* (London, 1972)

Duboy, Philippe, *Lequeu: An Architectural Enigma* (London, 1986)

Kluckert, Ehrenfried, *Vom Himmelsglobus zum Sternentheater* (Hamburg, 2005)

Lachièze-Rey, Marc, and Jean-Pierre Luminet, *Celestial Treasury: From the Music of the Spheres to the Conquest of Space* (Cambridge, 2001)

Lehni, Roger, *L'Horloge astronomique de la cathédrale de Strasbourg* (Saint-Ouen, 2011)

Schlee, Ernst, *Der Gottorfer Globus Herzog Friedrichs III* (Heide, 2002)

Smith, Earl Baldwin, *The Dome: A Study in the History of Ideas* (Princeton, NJ, 1933)

Voltaire, François-Marie Arouet, *The Princess of Babylon* (London, 1927)

Budge, E. A. Wallis, *The Gods of the Egyptians* (London, 1904)

Warmenhoven, Adrie, *Royal Eisinga Planetarium* (Franeker, 2000)

2 PLANETARY PROJECTION

Benjamin, Walter, *One-way Street and Other Writings* (London, 2009)

Füssl, Wilhelm, *Oskar von Miller* (Munich, 2005)

Gropius, Walter, *The Theatre of the Bauhaus* (Middletown, CT, 1972)

Krausse, Joachim, 'Architektur aus dem Geist der Projektion: Das Zeiss Planetarium', in *Wissen in Bewegung: 80 Jahre Zeiss-Planetarium*, ed. Hans-Christian von Herrmann (Jena, 2006), pp. 51–84

Letsch, H., *Das Zeiss Planetarium* (Jena, 1959)

Moholy-Nagy, László, *The New Vision, from Material to Architecture* (New York, 1932)

Simon, Joan, and Brigitte Leal, eds, *Alexander Calder: The Paris Years, 1926–33* (New York, 2008)
Villiger, Walter, *Das Zeiss Planetarium* (Jena, 1926)

3 RED STAR, WHITE STAR

Cleary, Richard, *Frank Lloyd Wright: From Within Outward* (New York, 2009)
Fisher, Clyde, *The Hayden Planetarium of the American Museum of Natural History* (New York, 1934)
Fox, Philip, *Adler Planetarium and Historical Collection* (Chicago, IL, 1933)
Grazia, Alfred de, ed., *The Velikovsky Affair* (London, 2006)
King, Henry C., *Geared to the Stars: The Evolution of Planetariums, Orreries and Astronomical Clocks* (Toronto, 1978)
Lewis, Ralph M., *Cosmic Mission Fulfilled* (San Jose, CA, 1978)
Marché, Jordan D., *Theatres of Time and Space* (New Brunswick, NJ, 2005)
Mayakovsky, Vladimir, *Polnoe sobranie sochinenii v dvenadtsati tomakh* (Moscow, 1941)
Millard, Doug, ed., *Cosmonauts: Birth of the Space Age* (London, 2015)
Rodchenko, Aleksandr, *Experiments for the Future: Diaries, Essays, Letters and Other Writings*, ed. Alexander N. Lavrentiev (New York 2005)
Schwarzman, Arnold, *Griffith Observatory: A Celebration of its Architectural Splendour* (Los Angeles, CA, 2014)
Young, George M., *The Russian Cosmists: The Esoteric Futurism of Nikolai Fedorov and His Followers* (Oxford, 2012)

4 OUTER PATHS

Ballard, J. G., *The Drowned World* (London, 1962)
Cornejo, Antonio, *Recuerdos de sus orígenes planetario de la ciudad de Buenos Aires* (Buenos Aires, 2015)
Cortázar, Julio, *Prosa del Observatorio* (Barcelona, 1972)
Griffiths, Alison, *Shivers Down Your Spine* (New York, 2008)
Herrmann, Dieter B., *Astronom in zwei Welten* (Berlin, 2008)
King, Henry C., *The London Planetarium* (London, 1962)
Koyré, Alexandre, *From the Closed World to the Infinite Universe* (Baltimore, MD, 1957)
Munro, Alice, *The Moons of Jupiter* (London, 1982)
Rich, Adrienne, *The Fact of a Doorframe: Poems, 1950–1984* (New York, 1984)
Sarraute, Nathalie, *Le Planetarium* (Paris, 1959)

5 VISIBLE, INVISIBLE

Close, Frank, *Nothing: A Very Short Introduction* (Oxford, 2015)
Coles, Peter, *Cosmology: A Very Short Introduction* (Oxford, 2001)
Devorkin, David H., and Robert W. Smith, *The Hubble Cosmos: Twenty-five Years of New Vistas in Space* (Washington, DC, 2015)
Futter, Ellen B., and Amy Weisser, *The Rose Center for Earth and Space: A Museum for the Twenty-first Century* (New York, 2001)
Green, Brian, *The Hidden Reality: Parallel Universes and the Deep Laws of the Cosmos* (London, 2011)
Keller, Coey, ed., *Brought to Light, Photography and the Invisible, 1840–1900* (New Haven, CT, 2009)
Kraupe, Thomas W., *'Denn was innen, das ist draussen': Die Geschichte des modernen Planetariums* (Hamburg, 2005)
Legro, Ron, and Avi Lank, *The Man Who Painted the Universe* (Madison, WI, 2015)
Mitchell, William J., *The Reconfigured Eye: Visual Truth in the Post-photographic Era* (Cambridge, MA, 1992)
Petersen, Carolyn, *Visions of the Cosmos* (Cambridge, 2003)
Wilford, John Noble, ed., *Cosmic Dispatches: The New York Times Reports on Astronomy and Cosmology* (New York, 2002)

ACKNOWLEDGEMENTS

The writing of this book has involved numerous conversations, discussions and research based on a wide range of sources.

From Reaktion Books I thank Vivian Constantinopoulos for supporting the proposal and her excellent and patient advice on the text and images, Amy Salter for editing, and the other members of the Reaktion Books team who made this book possible. In addition I thank Wolfgang Wimmer, Marte Schwabe, Dominique Schmied and Maria Bischoff at the Zeiss-Archiv in Jena for considerable assistance with illustrations; Sharon Shanks of *Planetarium* magazine; Charlotte Burford formerly of the Madame Tussauds Archives; Joachim Krausse; Tom Weaver, editor of the AA Files; Sina Najafi editor of *Cabinet* magazine; Tim Florian Horn of the Berlin Planetarium; Dieter B. Herrmann formerly of the Berlin Planetarium; Kzenia M Vozdigan and Evgenia M. Lupanova of the Kunstkamera St Petersburg; Felix Lühning; Gus Nielson and Beatrix Cordón de Jan; Andrie Warmenhoven of the Eisinga Planetarium; Mai Reitmeyer of the Hayden Planetarium; Gertrud Schille; Nadezda Gobova; Javier Ramirez of the Lima Planetarium; Julie Scott of the San Jose Planetarium; Jack van der Palen; Sridama Dasa of the Vedic Planetarium; Tom Kerss of the Peter Harrison Planetarium; A. R. Reshni; Hari Nadakumar; Touraj Daraee; Iain Ritchie; Simon Conolly; Tim MacFarlane; Mark C. Peterson of the extraordinary Loch Ness Productions planetarium database; Prof Matthias Ludwig of the Ulrich-Müther-Archiv in Wismar; Dionisios Simopoulos of the Athens Planetarium; Hari Nandakumar of Sri Sathya Sai Space Theatre; Chris Seale of Spitz inc; Frank Kovac of the Kovac Planetarium; and Jonathan Meades, Murray Fraser, Mark Dorrian, John Bold, Andrew Peckham, Dusan Decermic, John Hutchinson, Julian Krüger, Antje Buchholz and Andrew Higgott.

Special thanks to the Sabine Kunze and Milo Firebrace Kunze constellations. Earlier versions of sections of the text were published as *The Missing Planet* in AA Files 66 and as *Red Star Theatre* in *Cabinet* 57.

PHOTO ACKNOWLEDGEMENTS

The author and publishers wish to express their thanks to the below sources of illustrative material and/or permission to reproduce it. Every effort has been made to contact copyright holders; should there be any we have been unable to reach or to whom inaccurate acknowledgements have been made please contact the publishers, and full adjustments will be made to any subsequent printings.

Bibliothèque Nationale de France: pp. 40, 41; Antje Buchholz: p. 145; Buenos Aires Planetarium: p. 169; Cardcow: p. 113; Deutsches Museum, Munich: p. 55; Eise Eisinga Planetarium: pp. 28, 29; Fonds Bonnier. SIAF/Cité de l'architecture et du patrimoine/Archives d'architecture du XXe siècle, Paris: p. 47; Andrew Higgott: p. 163; iStockphoto: pp. 24–5 (Vold77); ITAR/TASS Photo Agency/ Alamy Stock Photo: p. 88; James Kirk: p. 6; Frank Kovac: pp. 206, 207; Kunstkamera, St Petersburg: pp. 33, 34; Library of Congress Prints and Photographs Division, Washington, DC: p. 108 (WPA Federal Art Project, 1939); London Planetarium Archive/Madame Tussauds: pp. 127, 149, 153; Joe Mamer Photography/Alamy Stock Photo: p. 136; Moholy-Nagy Foundation: p. 79; Jack van der Palen: p. 196; Planetario di Lima: p. 170; Ian Ritchie Architects: p. 186; Rosicrucian Planetarium, San Jose: p. 120; Royal Museums Greenwich: p. 12; Spitzinc: pp. 133, 134; Sri Sathya Sai Space Theatre: p. 200; Temple of the Vedic Planetarium: p. 199; To Scale: p. 203; Frank Lloyd Wright Foundation: p. 105; Zeiss Archiv: pp. 10, 53, 58, 59, 60, 65, 66, 67, 68, 70, 71, 72, 73, 85, 91, 98, 107, 109, 116, 137, 139, 141, 142, 144, 151, 159, 160, 174, 180, 182, 184–5, 208.

Alfred Gracombe has published the image on p. 188 online, Anufdo has published the image on p. 161 online, Avjoska has published the image on

p. 192 online under conditions imposed by a Creative Commons Attribution-Share Alike 3.0 Unported license; Luis Argerich has published the image on p. 164 online under conditions imposed by a Creative Commons Attribution 2.0 Generic license; Nagoya Taro (名古屋太郎) has published the image on p. 195 online under the Creative Commons Attribution-Share Alike 3.0 Unported, 2.5 Generic, 2.0 Generic and 1.0 Generic license. Readers are free to share – to copy, distribute and transmit these works – or to remix – to adapt these works under the following conditions: they must attribute the work(s) in the manner specified by the author or licensor (but not in any way that suggests that they endorse you or your use of the work(s)) and if they alter, transform, or build upon the work(s), they may distribute the resulting work(s) only under the same or similar licenses to those listed above).

INDEX

Page numbers in *italic* refer to illustrations